IEA Research for Education

A Series of In-depth Analyses Based on Data of the International Association for the Evaluation of Educational Achievement (IEA)

Volume 18

The International Association for the Evaluation of Educational Achievement (IEA) is an independent nongovernmental nonprofit cooperative of national research institutions and governmental research agencies that originated in Hamburg, Germany in 1958. For over 60 years, IEA has developed and conducted high-quality, large-scale comparative studies in education to support countries' efforts to engage in national strategies for educational monitoring and improvement.

IEA continues to promote capacity building and knowledge sharing to foster innovation and quality in education, proudly uniting more than 60 member institutions, with studies conducted in more than 100 countries worldwide.

IEA's comprehensive data provide an unparalleled longitudinal resource for researchers, and this series of in-depth peer-reviewed thematic reports can be used to shed light on critical questions concerning educational policies and educational research. The goal is to encourage international dialogue focusing on policy matters and technical evaluation procedures. The resulting debate integrates powerful conceptual frameworks, comprehensive datasets and rigorous analysis, thus enhancing understanding of diverse education systems worldwide.

Dirk Hastedt • Paulína Koršňáková
Branislav M. Ranđelović
Editors

Educating Pupils for Environmental Sustainability in Europe

Findings from TIMSS 2023 Data

Springer

Editors
Dirk Hastedt
IEA
Amsterdam, The Netherlands

Paulína Koršňáková
International Association for Evaluation of
Educational Achievements (IEA)
Amsterdam, The Netherlands

Branislav M. Ranđelović
University of Niš
Belgrade, Serbia

ISSN 2366-1631 ISSN 2366-164X (electronic)
IEA Research for Education
ISBN 978-3-032-14724-0 ISBN 978-3-032-14725-7 (eBook)
https://doi.org/10.1007/978-3-032-14725-7

This work was supported by International Association for the Evaluation of Educational Achievement.

Co-*Funded by the European Union. Views and opinions expressed are however those of the author(s) only and do not necessarily reflect those of the European Union or the European Education and Culture Executive Agency (EACEA). Neither the European Union nor EACEA can be held responsible for them.*

This Springer imprint is published by the registered company Springer Nature Switzerland AG
The registered company address is: Gewerbestrasse 11, 6330 Cham, Switzerland

Series Editors' Foreword

IEA's mission is to enhance knowledge about education systems worldwide and to provide high-quality data that will support education reform and lead to better teaching and learning in schools. In pursuit of this aim, it conducts, and reports on, major studies of student achievement in literacy, mathematics, science, citizenship, and digital literacy. These studies, most notably TIMSS, PIRLS, and ICCS, are well established and have set the benchmark for international comparative studies in education.

The studies have generated vast datasets encompassing student achievement, disaggregated in a variety of ways, along with a wealth of contextual information, which together contain considerable explanatory power. The numerous reports that have emerged from them are a valuable contribution to the corpus of educational research.

Valuable though these detailed reports are, IEA's goal of supporting education reform needs something more: a deep understanding of education systems and the many factors that bear on student learning, gained through in-depth analysis of IEA global datasets. IEA has long championed such analysis and facilitates scholars and policymakers in conducting secondary analysis of our datasets. So, we provide software such as the International Database Analyzer to encourage the analysis of our datasets and support numerous publications, including a peer-reviewed journal—*Large-scale Assessment in Education*—dedicated to the science of large-scale assessment and to publishing articles that draw on large-scale assessment databases. We also organize a biennial international research conference to nurture exchanges between researchers working with IEA data (https://www.iea.nl/our-conference).

The *IEA Research for Education* series represents a further effort by IEA to capitalize on our unique datasets, so as to provide powerful information for policymakers and researchers. Each report focuses on a specific topic and is produced by a dedicated team of leading scholars on the theme in question. Teams are selected on the basis of an open call for tenders; there are two such calls a year. Tenders are subject to a thorough review process, as are the reports produced (full details are available on the IEA website).

This report revisits the theme of environmental sustainability. A previous volume in the series, number 16 published in 2025, focused on students' knowledge and willingness to act pro-environmentally. The focus here is on learning for sustainability in the European context.

The report is framed within the European Union policy context for environmental sustainability learning and the Trends in International Mathematics and Science Study (TIMSS). The European sustainability competence framework, known as GreenComp, was published in 2022 and provides a comprehensive platform for scrutinizing learning for sustainability in schools. TIMSS is a large-scale international assessment that has measured student achievement in mathematics and science at grades four and eight every 4 years since 1995 and collects information on a wide range of contextual variables. The test battery provides comprehensive coverage of mathematics and science curricula, including many items relating to students' understanding of the environment; these range from biodiversity and food chains to deforestation, pollution, and natural hazards. An Environmental Awareness Scale based on these items was constructed for the 2019 and 2023 assessments.

The distinctive contribution of this report is its pulling together of perspectives, and empirical data, from TIMSS and GreenComp, alongside Sustainable Development Goal 4.7. Through mapping the GreenComp onto the TIMSS 2023 questionnaires and conducting insightful analyses of TIMSS 2023 data, it offers a detailed and richly contextualized account of European students' sustainable environmental learning and the factors associated with it. The country comparisons embedded within the study provide a valuable source of stimulus for reflection and action at national policy, school, and classroom levels.

University of Warwick
Coventry, UK
Seamus Hegarty

Indiana University
Bloomington, IN, USA
Leslie Rutkowski

Competing Interests The authors have no competing interests to declare that are relevant to the content of this manuscript.

Contents

1 **The European Sustainability Competence Framework (GreenComp): Introducing the Context for Investigating Learning for Environmental Sustainability** 1
Guia Bianchi, Ulrike Pisiotis, and Paulina Koršňáková

2 **Sustainability, Environment and Competencies to Act: Road Ahead for Science Education Across Europe** 27
Jelena Radišić and Erik Knain

3 **The European Sustainability Competence Framework (GreenComp) in Light of TIMSS 2023 Data** 53
Paulína Koršňáková, Tahira Ali Qadri, and Wangqiong Ye

4 **Attaining Knowledge on Environmental Sustainability: Knowledge and Understanding of Environmental Sustainability Revealed by TIMSS 2023 Data in European Education Systems** 87
Barbara Japelj Pavešić and Jure Novak

5 **Teaching Environmental Sustainability in the Science Classroom: Lessons from and for the TIMSS Environmental Framework** 139
Jelle Boeve-de Pauw, Michiel van Harskamp, Wanda Sass, Elisa Caponera, Riccardo Pietracci, and Laura Palmerio

6 **The Role of School Organizations in Promoting Sustainability** 165
Inés García-Bohórquez, Dries Verhelst, Fernando Martínez-Abad, and Camilo Ruiz

7 **Exploring the Link Between Sustainability Success and Student Knowledge and Attitudes: TIMSS 2023 Perspectives** 189
Svatava Janoušková, Paulina Koršňáková, Libor Klement, Michaela Janoušková, Michal Blaško, and Pavel Zikl

Chapter 1
The European Sustainability Competence Framework (GreenComp): Introducing the Context for Investigating Learning for Environmental Sustainability

Guia Bianchi, Ulrike Pisiotis, and Paulina Koršňáková

1.1 Introduction

Humankind has coined many terms to describe the current alarming state of planetary health. The discourse has evolved from describing complex and interrelated environmental and social challenges as "wicked problems" (Murgatroyd, 2010), to talking about broader concepts such as "Anthropocene" or "Technocene" (López-Corona & Magallanes-Guijón, 2020); or referring to a "triple planetary crisis" (United Nations Framework Convention on Climate Change [UNFCC], 2022) or "polycrisis" (Lawrence et al., 2024), and discussing the distress caused by environmental problems with concepts such as "eco-anxiety" (Coffey et al., 2021) or "solastalgia" (Albrecht et al., 2007). Terms are not scarce, nor is our knowledge about how humans are contributing to accelerating these problems and interconnected crises. For example, 2024 was the warmest year recorded in Europe, following the already exceptionally warm 2023, with average temperatures exceeding 1.5 °C above pre-industrial levels (Copernicus Climate Change Service, 2025; European Environment Agency, 2024).

G. Bianchi (✉)
Joint Research Centre, Seville Site, European Commission, Sevilla, Spain
e-mail: guia.bianchi@ec.europa.eu

U. Pisiotis
Directorate-General Education, Youth, Sports and Culture, European Commission, Brussels, Belgium
e-mail: ulrike.pisiotis@ec.europa.eu

P. Koršňáková
International Association for Evaluation of Educational Achievements (IEA), Amsterdam, The Netherlands
e-mail: p.korsnakova@iea.nl

D. Hastedt et al. (eds.), *Educating Pupils for Environmental Sustainability in Europe*, IEA Research for Education 18,
https://doi.org/10.1007/978-3-032-14725-7_1

Despite such information, moving from awareness and knowledge to taking action is not straightforward. Multiple factors contribute to this gap, and learning has a vital role to play in addressing it. Lifelong learning is necessary to promote lasting changes (Bianchi, 2020; United Nations Educational, Scientific and Cultural Organization [UNESCO], 2021) and advance sustainability transitions (Sterling, 2010). Learning for sustainability is therefore crucial for the present and future of our planet (Bianchi et al., 2022) to foster the kind of critical thinkers and ethical problem-solvers needed to promote a sustainable society (Sipos et al., 2008). Learning for sustainability is defined as learning that "aims to nurture a sustainability mindset from childhood to adulthood with the understanding that humans are part of and depend on nature. Learners are equipped with knowledge, skills and attitudes that help them become agents of change and contribute individually and collectively to shaping futures within planetary boundaries" (Bianchi et al., 2022, p. 13).

Moreover, the green transition is transforming economies, with profound changes in technology, production, services, consumption, and investment. Upskilling and reskilling are essential for Europe to maintain its competitiveness and thrive in a low-carbon economy. The European Commission's 2024 foresight report considers climate change and environmental degradation as "an existential threat to the European Union and to the world" (European Commission, Directorate-General for Research and Innovation, 2024, p. 3). It further describes the necessary transition towards greener and more sustainable economies as "a game changer in the [European Union] labor market alongside digitalization and automation" (p. 3) and notes that "skill needs will change with impacts far beyond the key occupations driving them, affecting all economic sectors" (p. 3).

The Union of Skills, the European Union's overarching strategy on skills development and lifelong learning, adopted in March 2025, seeks to address evolving skill needs and enhance Europe's competitiveness, including in achieving a successful green transition and circular economy (European Commission, 2025). The foundation of the Union of Skills is education and training as embodied in the European Education Area (EEA). This European Union policy and cooperation framework aims to build more resilient and inclusive national education and training systems among European Union Member States.

The EEA provides support by offering resources and tools to European Union countries and education stakeholders to support their efforts in improving teaching and learning. GreenComp, the European reference document on sustainability competences, published in January 2022, is one such resource, developed by the European Commission (Bianchi et al., 2022). It aligns with the Sustainable Development Goals (SDGs), particularly Target 4.7 of SDG 4 on quality education, in both objective and content.

This chapter first introduces the overall European Union approach to skills and education. It then describes competence-based education with particular emphasis on learning for sustainability and sustainability competences. Following this, the GreenComp framework, and how it has been applied by users since its publication, is presented. A brief introduction to the International Association for the Evaluation of Educational Achievement (IEA) Trends in International Mathematics and Science Study (TIMSS) provides a basis for discussion of whether and how TIMSS 2023 data can be mapped to GreenComp. Finally, an overview of the following chapters in this volume which review TIMSS 2023 data through the lens of learning for sustainability is presented, highlighting the current landscape of achievements in Europe and informing the future development of TIMSS questionnaires.

1.2 European Education Area (EEA)

In line with the Treaty on the Functioning of the European Union (European Union, 2012), education and training policies are the responsibility of the European Union Member States (Art. 165). The European Union's role is primarily supportive and complementary to the efforts of the Member States. It manifests itself through collaboration, the development of reference tools, and the sharing of best practices among Member States within the EEA or its predecessor, the Strategic Framework for European Cooperation in Education and Training. Taking an integrated approach, the EEA aims to: enhance synergies with other relevant initiatives, including the European Research Area and the Bologna Process; monitor progress towards achieving the EEA through relevant targets and indicators; and, foster the integration of education and training into the European Semester process, the European Union's economic and fiscal policies coordination framework. Cooperation between Member States in education and training is encouraged and implemented through the Open Method of Coordination to coordinate national policies and promote mutual learning, to address common challenges.

The European Union provides funding for education and training initiatives through various programmes—notably, Erasmus+, which supports student and staff exchanges, cooperation between educational institutions, and the development of innovative educational practices. The European Union further supports research and data collection to inform policy decisions and improve education and training systems, including studies conducted by the Eurydice network and other bodies.

The EEA is a shared vision of the 27 Member States of the European Union: Austria, Belgium, Bulgaria, Croatia, Cyprus, the Czech Republic, Denmark, Estonia, Finland, France, Germany, Greece, Hungary, Ireland, Italy, Latvia, Lithuania, Luxembourg, Malta, the Netherlands, Poland, Portugal, Romania, the Slovak Republic, Slovenia, Spain, and Sweden. One of the overarching dimensions of the EEA is supporting the association of the Western Balkans: Albania, Bosnia and Herzegovina, Kosovo,[1] Montenegro, the Republic of North Macedonia, and Serbia. This region will also be explored in this volume.

[1] This designation is without prejudice to positions on status, and is in line with UNSCR 1244/1999 (United Nations, 1999) and the International Court of Justice (ICJ) Opinion on the Kosovo declaration of independence (ICJ, 2010).

1.3 European Union Skills and Education Policy

The Draghi Report on the future of European competitiveness (Draghi, 2024) warned that without decisive action on productivity, investment and skills development, Europe risks losing its global competitiveness, particularly in strategic sectors such as clean technology, artificial intelligence and advanced manufacturing. According to the report, basic numeracy and literacy skills are crucial for a thriving and competitive economy but not enough to cope with and shape a rapidly evolving socioeconomic context. The report emphasizes that "society must develop awareness, practices and skills to function in a more sustainable and circular fashion" (p. 257), which would lead towards a green transition into a more resilient future.

Similarly, the European Commission 2024–2029 political guidelines stress the importance of enhancing skill levels across the European Union through lifelong learning by building a Union of Skills which will boost European competitiveness and align workforce capabilities with the needs of the green and digital economy (European Commission, 2024). The Union of Skills is intended to build on actions developed under the EEA (European Commission, 2025, p. 6), the European Union's policy and cooperation framework since 2020.

1.3.1 Competence-Based Education in Europe

Competence-based learning forms the basis for education and training policies in the European Union. It focuses on what is learned, rather than what has been taught, thus putting learners at the center. Competences comprise knowledge, skills and attitudes (Council of the European Union, 2018, pp. 1–13).

A competence-based education that helps learners develop sustainability skills based on knowledge and attitudes can help promote responsible action and stimulate willingness to take or demand action at local, national and global levels. Developing sustainability competences enables learners to overcome the cognitive dissonance that comes from knowing about an issue but lacking the intention, willingness or ability to act.

The European sustainability competence framework, GreenComp, both results from and reinforces the shift to competence-based education across Europe. European Union policy support for competence-based education can be traced back to 2006 when the Recommendation of the European Parliament and of the Council on key competences for lifelong learning (Council of the European Union, 2006) was adopted by European Union Member States. According to the Recommendation, "Competences are defined … as a combination of knowledge, skills and attitudes appropriate to the context. Key competences are those which all individuals need for personal fulfilment and development, active citizenship, social inclusion and employment" (Council of the European Union, 2006, p. 13).

At the time, eight key competences were defined which were updated in 2018 (Council of the European Union, 2018) to comprise the following:

- Literacy competence
- Multilingual competence
- Mathematical competence and competence in science, technology and engineering
- Digital competence
- Personal, social and learning to learn competence
- Citizenship competence
- Entrepreneurship competence
- Cultural awareness and expression competence

The adapted and revised version of the Recommendation, adopted by European Union Member States in 2018, maintained the original 2006 key competences in number (eight) but extended their scope to reflect changes in society, technology, and the labor market (Council of the European Union, 2018). For example, "communication in the mother tongue" with a focus on reading and writing became "literacy competence" with a broader meaning including critical thinking and media literacy. "Communication in foreign languages" became "multilingual competence," including the dimension of intercultural understanding. "Mathematical competence and *basic* [emphasis added] competences in science and technology" was changed to "mathematical competence and competence in science, technology and engineering." The "learning to learn" competence with a focus on independent learning was updated to "personal, social and learning to learn competence" and covers well-being and self-regulation.

The competences are considered equally important. Transversal skills such as critical thinking, problem solving, creativity, and communication are embedded throughout the key competences.

In some Member States, the 2006 Recommendation and the 2018 update initiated the shift from content- and knowledge-based education to competence-based education. In others, it strengthened this process already underway, which is why it is regarded as a catalyst "towards competence orientation" across Europe (European Commission, Directorate-General for Education, Youth, Sport and Culture, 2022, p. 10).

The time elapsed since the first Recommendation has allowed for clarification of what this shift towards a competence approach in teaching and learning means and what it does not. It does not mean a renunciation or even neglect of knowledge. Rather, knowledge is a set component of any competence and comprises "the facts and figures, concepts, ideas and theories which are already established and support the understanding of a certain area or subject" (Council of the European Union, 2018). At the same time, knowledge is not an end in itself but part of a wider competence that enables the learner to apply theoretical knowledge in practice and in other contexts. A competence-based education is thus both wider and deeper than a predominantly knowledge- and content-centered approach. Knowledge in such an understanding is complemented by skills, which are defined "as the ability and capacity to carry out processes and use the existing knowledge to achieve results." The third component is attitudes, which are described as "the disposition and mind-sets to act or react to ideas, persons or situations" (Council of the European Union, 2018). A competence-based approach to learning thus involves what has been described in other contexts—notably, related to learning for sustainability—as the head (knowledge), the heart (attitudes) and the hand (skills). This holistic way of addressing learning and the learner is an important prerequisite for learners to engage in learning for sustainability and develop sustainability competences (Sipos et al., 2008).

Similar to transversal skills such as critical thinking, sustainability was not referenced as a separate (key) competence in either the 2006 or the 2018 Recommendations. Instead, it was included as a cross-cutting concern that manifested itself particularly within the STEM (science, technology, engineering, and mathematics), citizenship and entrepreneurship competence. Thus, while duly acknowledged, sustainability could be perceived as secondary in relation to other competences due to competing subject aims and priorities within curricular documents.

This perceived undervaluation of sustainability competences is reflected in learning outcomes that lean towards content and knowledge but less so to skills and attitudes, thus not fully engaging in the holistic approach that is conducive to learning for sustainability (Sterling, 2001). A survey of more than 22,000 young people (15–35 years old) carried out in 23 European Union countries in October and November 2020 confirms that young people in Europe are very worried about climate change and its consequences (Dunne & Bijwaard, 2021). At a global scale, a study on climate anxiety which polled 10,000 young people (16–25 years old) in ten countries[2] found that nearly six in ten young people were very or extremely worried about climate change (Hickman et al., 2021).

Knowledge deficiencies or a lack of awareness is, however, not the reason for young people's eco-anxiety. The 2020 European Union survey confirms that climate deniers among young people are in fact a small minority; more than eight out of ten young Europeans (83%) consider humans and human activity to be the main cause of climate change (Dunne & Bijwaard, 2021). Results from the Organisation for Economic Co-operation and Development (OECD) PISA[3] 2018 cycle showed that 79% of 15-year-olds are fully aware of climate change and global warming (Schleicher, 2021). In contrast, only an average of 57% believe they can do something about climate change. In many European Union countries, including Austria, Estonia, Germany, Hungary, Latvia, Romania, the Slovak Republic, and Slovenia, that figure was lower than 50% (Schleicher, 2021). This perceived lack of empowerment and agency among young people (Schleicher, 2021), when it comes to addressing climate change and other sustainability-related challenges, leaves

[2] The study covered Australia, Brazil, Finland, France, India, Nigeria, the Philippines, Portugal, the United Kingdom and the United States of America.

[3] PISA = P*rogramme for International Student Assessment.*

education systems and schools with an important task—to support young people in developing competences and agency to tackle climate change and environmental degradation.

Policies and decisions at political and corporate levels, in conjunction with familial and peer influences, significantly shape young people's perception of their role in addressing the climate crisis. It might indeed be hard or impossible to delineate the precise shares influencing their experience and opinion. However, given that a large majority of young people finds themselves in school or education more generally, education systems and institutions have an important role to play for students to develop sustainability competences and take action. A recent pan-European study showed that all "European countries include the topic of sustainability in their curricula, and sustainability competences are cross-curricular in the majority of education systems" (European Commission, European Education and Culture Executive Agency, 2024, p. 10). Despite this, more needs to be done when it comes to teaching specific competences—notably, with regard to futures or systems thinking—and monitoring how learning for sustainability is implemented in schools (European Commission, European Education and Culture Executive Agency, 2024).

The strong link between sustainability competence and embedding learning for sustainability in teaching and learning is highlighted by different surveys. Results from IEA's 2022 International Civic and Citizenship Education Study (ICCS) include insights into students' attitudes and behaviors related to sustainability and environmental issues, which are part of the broader focus on global citizenship. A majority of students expressed strong concern about environmental issues, such as climate change, pollution, and loss of biodiversity. Countries that integrated environmental education into their civic and citizenship curricula tended to show higher levels of student awareness and engagement (Schulz et al., 2025).

A Flash Eurobarometer survey on the views of young people (aged 15–30) from 2019, revealed that many young people (41%) consider that school is failing to give them an adequate understanding of climate change, the environment, and how to live and act more sustainably (European Commission, Directorate-General for Education, Youth, Sport and Culture, 2019). Higher education students who participated in a 2021 survey by Students Organizing for Sustainability (SOS) International thought that tertiary education is the sector that has encouraged them the most so far to "think and act to help the environment and other people" (SOS International, 2021). However, only 26% of respondents thought that sustainable development has been covered in depth by their course (SOS International, 2021).

This is of concern as Europe cannot afford wasted talent or disengagement of young people when it comes to initiating and contributing to positive change in society. Involving students in meaningful ways in education and training requires learner-centered pedagogies and can ultimately contribute to developing learners' feelings of "ownership" towards sustainability (Gayford, 2008), which is crucial for positive action.

1.3.2 Learning for Sustainability

To define what sustainability competences are, it is first necessary to define sustainability and learning for sustainability.

Sustainability is often defined using the definition of sustainable development cited in the 1987 Brundtland Report, as development that "meets the needs of the present without compromising the ability of future generations to meet their own needs" (Brundtland, 1987, Chapter 2, para. 1). This definition is not only widely recognized but also has the advantage of being concise and intuitive; it defines in plain language what is, in reality, a complex concept.

In fact, sustainability can have different meanings in different contexts and different interpretations depending on factors such as time, location and groups of people involved (Bianchi, 2020; Jickling & Wals, 2012; Molderez & Ceulemans, 2018). Furthermore, although sustainability and sustainable development are often used interchangeably, they bear important conceptual differences. Sustainability is a long-term goal, while sustainable development refers to the means and pathways to achieve progress in sustainable ways (UNESCO, 2021). Sustainability involves respecting other species, taking care of the environment while making positive impacts, and being just and fair to members of societies. It is not, or not only, about finding processes or technologies to allow humans to live as usual; it requires a different mindset, one that can help address the problems humans have created.

Thus, sustainability can be understood as "prioritizing the needs of all life forms and of the planet by ensuring that human activity does not exceed planetary boundaries" (Bianchi et al., 2022, p. 12).

Learning for sustainability aims to empower learners to become positive agents of change, take the initiative and inspire other people to make positive impacts in society and in the economy. In other words, this type of learning goes beyond preparing learners for the labor market but enables them to become engaged and responsible citizens who contribute to a resilient society (OECD, 2018). Learning for sustainability is therefore associated with transformative learning, where the aim is not to transfer concepts to learners but to enable a third order learning by engaging both individuals (for example, teachers and learners), institutions (for example, schools) and the context in which they operate (for example, the local community) through a holistic approach.

Learning for sustainability can therefore be defined as "the holistic and interdisciplinary learning experiences that enable learners to embody sustainability values, vision, and mindset" (European Commission, Directorate-General for Education, Youth, Sport and Culture, 2024a, p. 9). Similarly, GreenComp defines learning for sustainability as learning that, "aims to nurture a sustainability mindset from childhood to adulthood with the understanding that humans are part of and depend on nature. Learners are equipped with knowledge, skills and attitudes that help them become agents of change and contribute individually and collectively to shaping futures within planetary boundaries" (Bianchi et al., 2022, p. 13).

Learning for sustainability focuses on global problems such as climate change or biodiversity loss. However, how these global problems are framed and addressed must be place-based, reflecting the context where learners live, operate, work, vote and contribute to society.

Learning for sustainability and education for sustainable development are interrelated and mutually reinforcing; both aim to enable learners to acquire the set of sustainability competences that are "needed to live more sustainably, in changing patterns of consumption and production, in embracing healthier lifestyles and in contributing—both individually and collectively—to a more sustainable economy and society" (Council of the European Union, 2022, p. 2).

1.3.3 Sustainability Competences

The importance of identifying and embedding sustainability competences in education and training for sustainability is well acknowledged in academic and wider public discourse. At the same time, the vocabulary used is neither clear nor consistent. In this chapter, the terminology used in documents by the European Commission is followed.

As noted earlier, competence goes beyond knowledge alone and embeds skills and attitudes. Knowing is complemented by the possibility of acting on this knowledge (skills) and having the intention to act (attitudes). In this context, skills are an important subset of competences. When it comes to sustainability in the context of the European Union, sustainability competences is a concept used more for education, whereas *green skills* is used for training, employment and the so-called green jobs (Bianchi, 2020). This distinction is particularly useful to measure and evaluate demand and supply of skills, and what skills are needed for green jobs. Yet, from a sustainability viewpoint, it is rather limiting. Sustainable professionals of today and tomorrow should not only perform a job in a more eco-friendly way—thus acquiring certain skills or green skills—but also need to know why they should be doing it and have the intention to favor a more green behavior over a non-green one if the aim is to fully embrace a sustainable economy and society. Systems thinking and futures literacy help reveal the bigger picture of why one action should be favored over another (for example, using a refillable coffee cup versus recycling a coffee paper cup).

Moving on to what sustainability competences are, in the literature there is some convergence over what competences people need to possess to contribute to sustainability problems and opportunities. Indeed, sustainability competences have been typically studied within higher education (Brundiers et al. 2021; Wiek et al., 2015), but less so in other contexts (Bianchi, 2020), such as in primary education (Vesterinen & Ratinen, 2024). It has therefore been recognized that a competence framework for sustainability for lifelong learning and adaptable to different learning settings was lacking (Bianchi, 2020), in line with European Union policy documents calling for its development.

1.3.4 GreenComp

Despite broad agreement among scholars and practitioners on the importance of embedding sustainability and sustainability competences into educational curricula, the literature has lacked a shared direction and common understanding of a competence framework for sustainability, and specifically one for lifelong learning (Bianchi, 2020).

The adoption of the European Green Deal in 2019 (European Commission, 2019), the European Commission's strategy to become the first climate-neutral continent, marked a collective shift by decision-makers to embed sustainability in all policies and provided momentum to develop effective policies and guidance on learning for sustainability, including GreenComp.

GreenComp is embedded in the Council Recommendation on learning for green transition and sustainable development adopted in June 2022 (Council of the European Union, 2022). It is also part of an ecosystem of competence frameworks that the Commission has published since 2013 and marks an intentional shift towards sustainability education and competences following the ambitions announced in the European Green Deal.

Previous competence frameworks stemmed from the European Union policy on key competences for lifelong learning. These include frameworks on digital competences, DigComp (Vuorikari et al., 2022); entrepreneurial competences, EntreComp (Bacigalupo et al., 2016); and personal, social and learning-to-learn competences, LifeComp (Sala et al., 2020).

The aim of GreenComp is to promote "a sustainability mindset by helping users develop the knowledge, skills and attitudes to think, plan and act with empathy, responsibility, and care for our planet" (Bianchi et al., 2022, p. 2). It seeks to foster a common understanding for learners and education and training professionals regarding sustainability competences. It therefore provides a reference framework for designing teaching and learning activities to develop sustainability competences.

1.3.4.1 GreenComp Development

GreenComp is the result of consensus building based on a mixed method research process. A diverse group of approximately 75 experts and stakeholders was consulted at different phases of the framework's development (Table 1.1) to obtain their feedback and progressively reach a consensus. The group included experts on sustainability education and lifelong learning from academia and research institutions, youth representatives, educators, policy representatives from European Union Member States, and nongovernmental organizations.

Table 1.1 Phases of the GreenComp development

Q3–Q4 2020	Scoping studies
Q1 2021	**First proposal**
April 2021	**Expert workshop**
Q2 2021	Feedback elaboration
Q3 2021	**Proposal update**
September 2021	**Expert workshop**
Q3–Q4 2021	**Framework refinement**
October 2021	**Stakeholder workshop**
January 2022	Publication

Source: Based on *Figure 1. Main steps to develop GreenComp* in Bianchi et al. (2022, p. 8)

1.3.4.2 GreenComp Structure

The goal of GreenComp is to empower learners towards action taking based on sustainability values to create different futures while embracing complexity. GreenComp is divided into four competence areas, each subdivided into three competences (Table 1.2). Taken together, these 12 competences (re)construct the definition of sustainability.

As part of the framework, its authors developed knowledge–skills–attitudes statements for each of the 12 competences. The aim was threefold. First, to describe the sustainability competences by elaborating on their knowledge, skills and attitudes components. Second, to further refine the competences by including input from the involved experts and users on these statements. Third, to make the framework more practice-oriented, by providing approximately 15 statements per competence.

The 12 competences in GreenComp are "interrelated and interconnected and should be treated as parts of a whole ... GreenComp implies that sustainability as a competence is made of 12 building blocks" (Bianchi et al., 2022, p. 15). To illustrate this, GreenComp uses the metaphor of pollinators: bees represent the competences related to Area 4, *acting for sustainability*; plants that will bloom flowers as Area 3, *envisioning sustainable futures*; the beehive as Area 1, *embodying sustainability values*; and pollen and nectar, connecting all these actors, as Area 2, *embracing complexity in sustainability*. This metaphor highlights the interdependencies of the 12 competences, showing that they are fundamental to the survival of the whole ecosystem (Fig 1.1).

1.3.4.3 GreenComp in Practice

Since its publication in January 2022, there has been tremendous interest in the framework. It is being used in teacher education programs, curriculum design, the development of learning material, schools, and Erasmus+ and Horizon Europe

Table 1.2 The 4 competence areas and 12 competences, and their definitions, in GreenComp

Area 1. Embodying sustainability values	Area 2. Embracing complexity in sustainability	Area 3. Envisioning sustainable futures	Area 4. Acting for sustainability
1.1 Valuing sustainability *To reflect on personal values, and identify and explain how values vary among people and over time, while critically evaluating how they align with sustainability values*	2.1 Systems thinking *To approach a sustainability problem from all sides; to consider time, space and context in order to understand how elements interact within and between systems*	3.1 Futures literacy *To envision alternative sustainable futures by imagining and developing alternative scenarios and identifying the steps needed to achieve a preferred sustainable future*	4.1 Political agency *To navigate the political system, identify political responsibility and accountability for unsustainable behavior, and demand effective policies for sustainability*
1.2 Supporting fairness *To support equity and justice for current and future generations and learn from previous generations for sustainability*	2.2 Critical thinking *To assess information and arguments, identify assumptions, challenge the status quo, and reflect on how personal, social and cultural backgrounds influence thinking and conclusions*	3.2 Adaptability *To manage transitions and challenges in complex sustainability situations and make decisions related to the future in the face of uncertainty, ambiguity and risk*	4.2 Collective action *To act for change in collaboration with others*
1.3 Promoting nature *To acknowledge that humans are part of nature; and to respect the needs and rights of other species and of nature itself in order to restore and regenerate healthy and resilient ecosystems*	2.3 Problem framing *To formulate current or potential challenges as a sustainability problem in terms of difficulty, people involved, time and geographical scope, in order to identify suitable approaches to anticipating and preventing problems, and to mitigating and adapting to already existing problems*	3.3 Exploratory thinking *To adopt a relational way of thinking by exploring and linking different disciplines, using creativity and experimentation with novel ideas or methods*	4.3 Individual initiative *To identify own potential for sustainability and to actively contribute to improving prospects for the community and the planet*

Source: Based on *Table 1. GreenComp areas, competences, and descriptors* in Bianchi et al. (2022, pp. 14–15)

projects. A report published by the European Commission in 2024 traced the uptake of the framework across education sectors and analyzed in 12 in-depth case studies how GreenComp was adapted to specific contexts and needs, what challenges arose when applying it, and what lessons are to be learned from those cases (European Commission, Directorate-General for Education, Youth, Sport and Culture, 2024b).

Fig. 1.1 Visual representation of GreenComp. Source: Reproduction of *Figure 3. Visual representation of GreenComp* from Bianchi et al. (2022, p. 16)

Another 2024 study (European Commission, European Education and Culture Executive Agency, 2024) mapped how green competences are integrated in "national regulations or other official top-level education documents, such as curricula, guidelines or similar steering documents" (p. 23) in primary and general secondary education in 39 European countries. Elements of systems thinking were found in most school curricula, when looking at the "interaction between humans and nature across space and time" (p. 37). However, regarding the core components of systems thinking identified in GreenComp—such as synthesis, emergence, interconnectedness, feedback loops or cascade effects—no curricula examples could be found (p. 37). Similarly, and even with a broader definition of futures literacy, the study found that this "competence is the least present in European curricula" (p. 39).

As of July 2025, Google Scholar indexed 653 citations of the GreenComp framework in academic articles, book chapters and monographs, excluding translated versions of the report.

To support the uptake of GreenComp, the European Commission has established a community of practice on the Education for Climate Coalition platform, providing a forum for exchange, resources and sharing of knowledge.[4]

[4] Accessible at: https://education-for-climate.ec.europa.eu/community/GreenCompCommunity

1.4 Introduction to the TIMSS Assessment of Science Education and Environmental Awareness

TIMSS is a large-scale international student assessment that has been conducted every four years since 1995. It evaluates the mathematics and science achievement of students in the fourth and eighth grades across participating education systems and over time allows for monitoring trends in student achievement and the context of learning.

From the beginning, TIMSS adopted a curricular model (see Chap. 4) connecting the assessment content with students' opportunities to learn in school and ensuring that the test measures how well students have mastered the material taught in their national curricula.

To ensure trend comparability, TIMSS has consistently aimed for a balanced representation of the three science content domains—life science, physical science, and Earth science—and the three cognitive skills domains—knowing, applying, and reasoning. While at the policy level countries experienced a shift from content- and knowledge-based education to competence-based education over time (European Commission, 2012), applying and reasoning were always part of the TIMSS framework in addition to the third cognitive domain of knowing. The balance has remained relatively stable since 1995. For example, in TIMSS 2023, the target distribution for fourth grade science content domains was 45% for life science, 35% for physical science, and 20% for Earth science; and for cognitive skills domains, 40% for knowing, 40% for applying, and 20% for reasoning (Mullis et al., 2021). The only change since TIMSS 2003 occurred in the science cognitive domains with 5% fewer items in applying and 5% more items in reasoning (Mullis et al., 2003).

Each domain includes specific topic areas that are age-appropriate for fourth- and eighth-grade students. Environmental topics have been integrated across the science content domains in ways that reflect both natural systems and human impact and connect scientific principles to real-world environmental challenges.

Environmental content in life science (or biology for the grade eight students) typically includes topics assessing students' understanding of the interconnectedness of life and how organisms interact with each other and their environment (ecosystems and food chains) and exploring how species survive in different habitats and the importance of biodiversity (adaptations and biodiversity). It also includes topics such as deforestation, pollution, and conservation efforts (resources and human impact on living systems).

Earth science covers environmental content related to Earth's systems (water cycles, soil, and atmospheric processes), natural hazards (earthquakes, floods, and their environmental consequences), weather and climate, and natural resources.

The availability of items assessing students' understanding of the environment was reviewed by the TIMSS & PIRLS International Study Center to construct an environmental awareness scale based on a wide range of topics specified in the TIMSS 2019 science assessment framework (Mullis & Martin, 2017). The TIMSS 2019 environmental awareness scale was based on 33 environmental items at the fourth grade and 41 environmental items at the eighth grade (for further details, see Yin & Foy, 2021). TIMSS 2023 built on this foundation by supplementing the TIMSS 2023 assessment framework with the TIMSS 2023 Environmental Attitudes and Behaviors Framework (Reynolds & Komakhidze, 2022) outlining how students' attitudes toward the natural environment, as well as their enactment of environmentally responsible behaviors, parents' behaviors, school policies, and teaching practices will be represented in TIMSS 2023 student, home, school, and teacher questionnaires.

In all, 72 education systems participated in TIMSS 2023. Of these, 23 are from the EEA and 6 are from the Western Balkan area (Table 1.3). All 6 of the Western Balkan participants are candidates for European Union accession. Out of the 23

Table 1.3 EEA and Western Balkan education systems in TIMSS 2023 with grade four and eight datasets

TIMSS 2023 participant	Grade four	Grade eight	EEA	Western Balkan
Albania	●			●
Austria		●	●	
Belgium (Flemish)	●		●	
Belgium (French)	●		●	
Bosnia and Herzegovina	●			●
Bulgaria	●		●	
Cyprus	●	●	●	
Czech Republic	●	●	●	
Denmark	●		●	
Finland	●	●	●	
France	●	●	●	
Germany	●		●	
Hungary	●	●	●	
Ireland	●	●	●	
Italy	●	●	●	
Kosovo	●			●
Latvia	●		●	
Lithuania	●	●	●	
Malta	●	●	●	
Montenegro	●			●
Netherlands	●		●	
N. Macedonia, Rep. of	●			●
Poland	●		●	
Portugal	●	●	●	
Romania	●	●	●	
Serbia	●			●
Slovak Republic	●		●	
Slovenia	●		●	
Spain	●		●	
Sweden	●	●	●	

education systems from the EEA, only 1 (Austria) did not collect data for the grade four population while 10 did not collect data for the grade eight population. The 6 Western Balkan education systems collected data only for the grade four population.

This available internationally comparable data will be used across the following chapters. Not all chapters will use all the data available and most of the chapters will focus on the grade four data representing 29 education systems (22 European Union participants, with Belgium covered by two education systems, Flemish and French, and 6 Western Balkan participants). Grade four data provide better coverage of the education area targeted by this publication. In addition, this data includes some information collected from parents, which improves data coverage within the targeted area. It is important to note, however, that grade eight data were already examined in a previous volume in this series focusing on willingness to act pro-environmentally based on TIMSS 2019 and ICCS 2016 (Isac et al., 2025).

1.5 GreenComp Representation in TIMSS 2023 Questionnaires

TIMSS 2023 questionnaires were developed based on the TIMSS 2023 framework which was complemented by the TIMSS 2023 Environmental Attitudes and Behaviors Framework referenced in the previous section. While Chap. 2 will provide further insights into how these frameworks relate to other theoretical frameworks linked to science, environmental and sustainability education (including GreenComp), this section will focus on the approach to mapping GreenComp using the TIMSS questionnaire data, explaining the item selection process and noting the choices made to reach the results. The results of the mapping will be used in Chap. 3, which analyzes TIMSS 2023 data and creates a baseline for measuring education's critical role in preparing learners to acquire the knowledge, skills, values, and attitudes needed to address climate change and promote sustainable ways of living.

Since GreenComp and the TIMSS 2023 frameworks were developed independently, the available TIMSS scales measuring factors or constructs do not directly relate to the GreenComp areas or competence definitions. For example, TIMSS 2023 adapted the two-dimensional Model of Ecological Values, based on the Theory of Ecological Attitude, by including two broader dimensions of environmental attitudes: preservation and utilization (Reynolds & Komakhidze, 2022). Some items linked to preservation corresponded with GreenComp Area 1, *embodying sustainability values* (in particular, 1.3, *promoting nature*), while others correspond with Area 4, *acting for sustainability* (in particular, 4.2 or 4.3, depending on whether they relate to a collective action or individual initiative).

However, all TIMSS items were developed to collect useful information or reflect current research. This allows for use of the more granular data behind the singular items used to measure students' environmental attitudes and environmentally responsible behaviors (student questionnaire); parenting practices embedded in parents' environmental attitudes and behaviors (home questionnaire); and teachers' attitudes, competence, and practices shaping the classroom context (teacher questionnaire) as outlined in the TIMSS 2023 Environmental Attitudes and Behaviors Framework (Reynolds & Komakhidze, 2022).

GreenComp areas outlined by the competence descriptors are further supported by statements describing knowledge, skills, and attitudes (see Appendix 2 in Bianchi et al., 2022), which provide deeper guidance and aid understanding of higher levels of the framework. On this basis, the present analysis focused on three of the four available questionnaires linked to the grade four population, collecting responses from students, their parents, and teachers. While students', parents' and teachers' responses offered input at a personal level—providing a glimpse of their attitudes, values, and activities—school principals were only asked for information pertinent to schools' emphasis on environmental sustainability within school policies and practices.

The first step involved selecting TIMSS 2023 questions containing relevant items. Full wording of the selected items, embedded within the questions from which they were drawn, is available in the chapter appendix. Next, each selected item was attributed to one of the four GreenComp areas, and the most appropriate competence was noted. Results were then compared and differences reconciled.

One example that illustrates the challenges and choices involved in the mapping comes from the home questionnaire (Table 1.4). Question 9 asked, "How often do you or someone else in your home do these things with your child?" The question included five items, with the response categories *every day*, *almost every day*, *sometimes*, and *never*. Several items related to more than one GreenComp area,

Table 1.4 GreenComp mapped to the TIMSS 2023 home questionnaire, Question 9

TIMSS home questionnaire, Q9 items	GreenComp area (competence)
(a) Discuss environmental problems	Area 2. Embracing complexity in sustainability (within the 2.2 Critical thinking, but also 2.1 System thinking or 2.3 Problem framing)
(b) Read or watch information about environmental problems	Area 4. Acting for sustainability (4.3 Individual initiative) and Area 2. Embracing complexity in sustainability (2.2 Critical thinking)
(c) Show him/her how to use less resources to help the natural environment	Area 4. Acting for sustainability (4.3 Individual initiative) and Area 1. Embodying sustainability values (1.3 Promoting nature)
(d) Spend time in nature	Area 1. Embodying sustainability values (1.3 Promoting nature)
(e) Encourage him/her to take action to protect the natural environment	Area 4. Acting for sustainability (4.3 Individual initiative) and Area 1. Embodying sustainability values (1.3 Promoting nature)

requiring decisions about which alignment to prioritize. For example, while the item "Read or watch information about environmental problems" could involve cognitive processes as described in Area 2, their presentation could only happen during discussions on environmental problems (Area 1). Nevertheless, this is an activity based on personal choice (individual initiative that could be linked to political agency) more aligned with GreenComp Area 4 than Area 1. In addition, it was not always possible to pinpoint one specific competence within some areas, so mapping remained on the level of areas, with some understanding on the coverage of their competence descriptors (Table 1.4).

Ultimately, not all GreenComp areas were covered by the items (specifically, there was no coverage of Area 3), meaning it was not possible to create a comprehensive GreenComp representation using TIMSS 2023 data from parents.

The teacher questionnaire provided more suitable items (ten items compared to the five in the home questionnaire) and covered all of the GreenComp areas. Area 3, *envisioning sustainable futures*, was linked to the item "Having students predict the outcomes of experiments or investigations" through competence descriptor 3.3, *exploratory thinking*. Here, however, a novel challenge was presented: While parents were acting as they pleased (there is no curriculum parents are supposed to obey), it was not clear from the teachers' responses whether they could be considered as a record of their personal characteristic, or compliance with the curricular requirements.

Finally, GreenComp representation was created using the student questionnaire, establishing 10 connection points provided by three items in Area 1, two in Area 2, one in Area 3, and four in Area 4 (Table 1.5). These results will be used in the Chap. 3 analysis.

1.6 Learning for Sustainability in This Volume

The following chapters of the volume provide several views on sustainability using different sets of TIMSS 2023 data. Chapter 2 focuses on important discussions regarding sustainability and explores how the concepts of sustainability and environmental sustainability competence intertwine with science education. It addresses efforts to measure environmental sustainability, providing an overview of the landscape for sustainability skills assessment created by three frameworks: the European sustainability competence framework (GreenComp), an environmental sustainability competence toolbox, and the TIMSS 2023 Environmental Attitudes and Behaviours Framework. Using the grade four data from TIMSS 2023, Chap. 3 employs latent class analysis to explore patterns in students' ecological engagement and examines variations in these patterns across gender and education systems from the European Union and Western Balkan participants in TIMSS 2023. It is based on the GreenComp representation in the TIMSS 2023 questionnaires, as outlined in this chapter, and provides a baseline assessment of environmental sustainability engagement among the assessed education systems. Chapter 4 explores the wealth

Table 1.5 GreenComp mapped to the TIMSS 2023 grade four student questionnaire

GreenComp area	Competence	Student questionnaire item code (as available in TIMSS database)	Item text
1	1.3 Promoting nature	ASBG11A	I care about the protection of plants and animals
1	1.3 Promoting nature	ASBG11C	I enjoy finding out what kinds of plants and animals live in my surrounding area
1	1.3 Promoting nature	ASBG11D	I enjoy being in nature
1	1.3 Promoting nature	ASBG12F	I tell my friends when they are doing things that harm the environment
2	2.2 Critical thinking	ASBS08H	My teacher asks me to show what I have learned
2	2.3 Problem framing	ASBS07G	Science teaches me how things in the world work
3	3.3 Exploratory thinking	ASBS06	In science lessons, how often does your teacher ask you to conduct science experiments?
4	4.3 Individual initiative	ASBG12A	I try to reuse things
4	4.3 Individual initiative	ASBG12B	I try to use less resources
4	4.2 Collective action	ASBG12E	I try to participate in group activities to help the environment

of TIMSS 2023 data at three levels of analysis. The first level involves the comparison of prescribed content in national curricula, outlining opportunities for students to learn about sustainability. The second level focuses on how extensively these topics are taught in classrooms. The third level assesses students' knowledge, awareness, and understanding of environmental sustainability. Chapter 5 examines how environmental sustainability is taught in primary science classrooms. Recognizing that teaching is a highly contextualized activity, the authors highlight two education systems—Italy and the Netherlands. Chapter 6 investigates school organizational factors in effective education for sustainable development within primary schools, emphasizing the whole school approach as a key enabler of sustainable change. The chapter discusses the implications of its findings for restructuring school organizational practices to align them with sustainability education policies, such as GreenComp and SDG 4, providing input for future assessments to better capture the holistic and process-oriented nature of the whole school approach. Finally, Chap. 7 combines TIMSS 2023 data with information obtained in the Sustainable Development Report 2024 (Sachs et al., 2024), which provides details on responsible production and consumption, waste and wastewater management, biodiversity protection, and other environmental sustainability issues in communities within countries. This chapter creates a more comprehensive picture of where students live and what can affect their worldview (knowledge, values, and attitudes). While all of the chapters are unique in their approach, data use, and limitations, they point towards similar conclusions that challenge expectations.

Appendix: Overview of the Questions from Different TIMSS 2023 Questionnaires Aligning with GreenComp, with the Selected Items Shown in Bold

Student Questionnaire

G11

How much do you agree with these statements? Fill one circle for each line.
Agree a lot/Agree a little/Disagree a little/Disagree a lot

(a) **I care about the protection of plants and animals**
(b) It makes me sad when nature is destroyed
(c) **I enjoy finding out what kinds of plants and animals live in my surrounding area**
(d) **I enjoy being in nature (e.g., <forests, parks, deserts>)**
(e) Addressing climate change should be a high priority

G12

How often do you do these things to help the natural environment?
Fill one circle for each line.
Every day/Almost every day/Sometimes/Never

(a) **I try to reuse things (e.g., <bags, bottles>)**
(b) **I try to use less resources (e.g., <water, food>)**
(c) I talk about how to help the environment (e.g., <saving water, picking up trash>)
(d) I try to learn about environmental problems (e.g., <climate change, endangered animals>)
(e) **I try to participate in group activities to help the environment**
(f) **I tell my friends when they are doing things that harm the environment**

MS6

In science lessons, how often does your teacher ask you to conduct science experiments?

Fill one circle only.
At least once a week/Once or twice a month/A few times a year/Never
MS7
How much do you agree with these statements about learning science?

Fill one circle for each line.
Agree a lot/Agree a little/Disagree a little/Disagree a lot

(a) I enjoy learning science
(b) I wish I did not have to study science
(c) Science is boring
(d) I learn many interesting things in science
(e) I like science
(f) I look forward to learning science in school
(g) **Science teaches me how things in the world work**
(h) I like to do science experiments
(i) Science is one of my favorite subjects

MS8

How much do you agree with these statements about your science lessons?
Fill one circle for each line.
Agree a lot/Agree a little/Disagree a little/Disagree a lot

(a) My teacher makes it clear what we should learn in each lesson
(b) My teacher is easy to understand
(c) My teacher has clear answers to my questions
(d) My teacher is good at explaining science
(e) My teacher does a variety of things to help us learn
(f) My teacher explains a topic again when we don't understand
(g) My teacher gives me helpful feedback on my work
(h) **My teacher asks me to show what I have learned**
(i) My teacher asks me to explain my answers

Home Questionnaire

9

How often do you or someone else in your home do these things with your child?
Check one circle for each line.
Every day/Almost every day/Sometimes/Never

(a) **Discuss environmental problems (e.g., <climate change, endangered animals>)**
(b) **Read or watch information about environmental problems (e.g., <climate change, endangered animals>)**
(c) **Show him/her how to use less resources (e.g., <water, energy>) to help the natural environment**
(d) **Spend time in nature (e.g., <forest, parks, deserts>)**
(e) **Encourage him/her to take action to protect the natural environment**

Teacher Questionnaire

TS2

In teaching science to the students in this class, how often do you ask them to do the following?

Check one circle for each line.

Every or almost every lesson/About half the lessons/Some lessons/Never

(a) Listen to me explain new science content
(b) Observe natural phenomena such as the weather or a plant growing and describe what they see
(c) Watch me demonstrate an experiment or investigation
(d) Read their textbooks or other resource materials
(e) Memorize facts and principles
(f) **Do field work outside the class**
(g) Work in mixed ability groups
(h) Work in same ability groups

TS3

How much emphasis do you place on the following when teaching science to students in this class?

Check one circle for each line.

A lot/Some/None

(a) **Encouraging students to ask questions about scientific phenomena**
(b) **Having students predict the outcomes of experiments or investigations**
(c) Having students create representations (e.g., models, graphs) to explain scientific phenomena
(d) Having students use scientific concepts to explain phenomena
(e) **Having students conduct experiments (hands-on or virtually)**

TS4

How often do you do the following when teaching this class?

Check one circle for each line.

At least once a week/Once or twice a month/A few times a year/Never or almost never

(a) **Develop students' positive attitudes toward the natural environment**
(b) Encourage students to use less resources (e.g., <water, energy>)
(c) **Discuss how student actions in and outside of school can help the natural environment**
(d) Discuss environmental issues (e.g., <climate change, endangered animals>)

TS5

Do you do these things to teach students about environmental issues and sustainability? Check one circle for each line.

Yes/No

(a) **Take students to visit natural areas (e.g., <a pond or meadow>)**
(b) **Have students participate in environmentally responsible activities (e.g., <pick up trash>)**
(c) **Have students do research or projects on a particular environmental topic (e.g., <pollution, climate change>)**
(d) **Provide opportunities for students to participate in outdoor environmental education programs outside of school**

References

Albrecht, G., Sartore, G. M., Connor, L., Higginbotham, N., Freeman, S., Kelly, B., Stain, H., Tonna, A., & Pollard, G. (2007). Solastalgia: The distress caused by environmental change. *Australasian Psychiatry, 15*(sup1), S95–S98. https://doi.org/10.1080/10398560701701288

Bacigalupo, M., Kampylis, P., Punie, Y., & Van den Brande, G. (2016). *EntreComp: The entrepreneurship competence framework.* European Commission: Joint Research Centre. Publications Office of the European Union. https://doi.org/10.2791/160811

Bianchi, G. (2020). *Sustainability competences. A systematic literature review.* European Commission: Joint Research Centre: Publications Office of the European Union. https://data.europa.eu/doi/10.2760/200956

Bianchi, G., Pisiotis, U., & Cabrera, M. (2022). *GreenComp. The European sustainability competence framework.* European Commission: Joint Research Centre. Publications Office of the European Union. https://doi.org/10.2760/13286

Brundiers, K., Barth, M., Cebrián, G., Cohen, M., Diaz, L., Doucette-Remington, S., Dripps, W., Habron, G., Harré, N., Jarchow, M., Losch, K., Michel, J., Mochizuki, Y., Rieckmann, M., Parnell, R., Walker, P., & Zint, M. (2021). Key competencies in sustainability in higher education—Toward an agreed-upon reference framework. *Sustainability Science, 16*(1), 13–29. https://doi.org/10.1007/s11625-020-00838-2

Brundtland, G. H. (1987). *Our common future: Report of the World Commission on Environment and Development.* United Nations General Assembly document A/42/427.

Coffey, Y., Bhullar, N., Durkin, J., Islam, M. S., & Usher, K. (2021). Understanding eco-anxiety: A systematic scoping review of current literature and identified knowledge gaps. *The Journal of Climate Change and Health, 3*, 100047. https://doi.org/10.1016/j.joclim.2021.100047

Copernicus Climate Change Service. (2025). *Global climate highlights 2024.* https://climate.copernicus.eu/global-climate-highlights-2024

Council of the European Union. (2006). Recommendation of the European Parliament and of the Council of 18 December 2006 on key competences for lifelong learning. *Official Journal of the European Union,* L 394/10–18. https://eur-lex.europa.eu/eli/reco/2006/962/oj/eng.

Council of the European Union. (2018). Council Recommendation of 22 May 2018 on key competences for lifelong learning. *Official Journal of the European Union,* C 189/1–13. https://eur-lex.europa.eu/legal-content/EN/TXT/?uri=oj:JOC_2018_189_R_0001.

Council of the European Union. (2022). Council Recommendation of 16 June 2022 on learning for the green transition and sustainable development. *Official Journal of the European Union,* C 243/1–9. https://eur-lex.europa.eu/legal-content/EN/TXT/?uri=oj:JOC_2022_243_R_0001.

Draghi, M. (2024). *The future of European competitiveness. Part B. In-depth analysis and recommendations.* European Commission. Publications Office of the European Union. https://commission.europa.eu/topics/eu-competitiveness/draghi-report_en#paragraph_47059

Dunne A., & Bijwaard D. (2021). *Pan-European survey: Main multi-country report.* Ipsos. https://eeb.org/wp-content/uploads/2021/04/IPSOS-Multi-Country-Report-complete.FINAL_.pdf.

European Commission. (2012). *Assessment of key competences in initial education and training: Policy guidance.* SWD/2012/0371 final. https://eur-lex.europa.eu/legal-content/EN/TXT/?uri=CELEX%3A52012SC0371.

European Commission. (2019). The European Green Deal. COM(2019) 640 final. https://eur-lex.europa.eu/legal-content/EN/TXT/?uri=celex:52019DC0640.

European Commission. (2024). *Political guidelines for the next European Commission 2024–2029.* https://commission.europa.eu/document/download/e6cd4328-673c-4e7a-8683-f63ffb2cf648_en.

European Commission. (2025). The union of skills. COM(2025) 90 final. https://eur-lex.europa.eu/legal-content/EN/TXT/?uri=celex:52025DC0090.

European Commission, Directorate-General for Education, Youth, Sport and Culture. (2019). *How do we build a stronger, more united Europe? The views of young people.* Flash Eurobarometer 478. Publications Office of the European Union. https://data.europa.eu/doi/10.2766/53982.

European Commission, Directorate-General for Education, Youth, Sport and Culture. (2022) *Key competences for all. Policy design and implementation in European school education. Final report.* Publications Office of the European Union. https://data.europa.eu/doi/10.2766/1951.

European Commission, Directorate-General for Education, Youth, Sport and Culture. (2024a). *Education and training monitor 2024. Comparative report.* Publications Office of the European Union. https://data.europa.eu/doi/10.2766/815875.

European Commission, Directorate-General for Education, Youth, Sport and Culture. (2024b). *GreenComp in practice. Case studies on the use of the European competence framework. Analytical report.* Publications Office of the European Union. https://data.europa.eu/doi/10.2766/053738.

European Commission, Directorate-General for Research and Innovation. (2024). *Futures of green skills and jobs in Europe in 2050. Scenarios and policy implications.* Publications Office of the European Union. https://data.europa.eu/doi/10.2777/36430.

European Commission, European Education and Culture Executive Agency. (2024). *Learning for sustainability in Europe. Building competences and supporting teachers and schools. Eurydice report.* Publications Office of the European Union. https://data.europa.eu/doi/10.2797/81397.

European Environment Agency. (2024). *European climate risk assessment.* Publications Office of the European Union. https://data.europa.eu/doi/10.2800/8671471.

European Union. (2012). Consolidated version of the Treaty on the Functioning of the European Union. *Official Journal of the European Union,* C 326/47–390. https://eur-lex.europa.eu/LexUriServ/LexUriServ.do?uri=CELEX:12012E/TXT:en:PDF.

Gayford, C. (2008). *Learning for sustainability: From the pupils' perspective.* World Wide Fund for Nature.

Hickman, C., Marks, E., Pihkala, P., Clayton, S., Lewandowski, E., Mayall, E., Wray, B., Mellor, C., & van Susteren, L. (2021). Climate anxiety in children and young people and their beliefs about government responses to climate change: A global survey. *The Lancet, Planetary Health, 5*(12), e863–e873. https://doi.org/10.1016/S2542-5196(21)00278-3

ICJ. (2010). *Accordance with International Law of the Unilateral Declaration of Independence in respect of Kosovo, Advisory Opinion, I.C.J. Reports 2010.* International Court of Justice. https://www.icj-cij.org/public/files/case-related/141/141-20100722-ADV-01-00-EN.pdf.

Isac, M. M., Sandoval-Hernández, A., & Sass, W. (Eds.). (2025). *Knowledge and willingness to act pro-environmentally. Perspectives from IEA TIMSS 2019 and ICCS 2016 data* (IEA Research for Education, Vol. 16). Springer. https://doi.org/10.1007/978-3-031-76033-4.

Jickling, B., & Wals, A. (2012). Debating education for sustainable development 20 years after Rio: A conversation between Bob Jickling and Arjen Wals. *Journal of Education for Sustainable Development, 6*(1), 49–57. https://doi.org/10.1177/0973408211006001111

Lawrence, M., Homer-Dixon, T., Janzwood, S., Rockstöm, J., Renn, O., & Donges, J. F. (2024). Global polycrisis: The causal mechanisms of crisis entanglement. *Global Sustainability, 7*(e6). https://doi.org/10.1017/sus.2024.1

López-Corona, O., & Magallanes-Guijón, G. (2020). It is not an Anthropocene; it is really the Technocene: Names matter in decision making under planetary crisis. *Frontiers in Ecology and Evolution, 8*, 214. https://doi.org/10.3389/fevo.2020.00214

Molderez, I., & Ceulemans, K. (2018). The power of art to foster systems thinking, one of the key competencies of education for sustainable development. *Journal of Cleaner Production, 186*, 758–770. https://doi.org/10.1016/j.jclepro.2018.03.120

Mullis, I. V. S., & Martin, M. O. (Eds.). (2017). *TIMSS 2019 assessment frameworks*. TIMSS & PIRLS International Study Center, Boston College. http://timssandpirls.bc.edu/timss2019/frameworks/

Mullis, I. V. S., Martin, M. O., Smith, T. A., Garden, R. A., Gregory, K. D., Gonzalez, E. J., Chrostowski, S. J., & O'Connor, K. M. (Eds.). (2003). *TIMSS assessment frameworks and specifications 2003* (2nd ed.). TIMSS & PIRLS International Study Center, Boston College. https://timssandpirls.bc.edu/timss2003i/frameworksD.html

Mullis, I. V. S., Martin, M. O., & von Davier, M. (Eds.). (2021). *TIMSS 2023 assessment frameworks*. TIMSS & PIRLS International Study Center, Boston College. https://timssandpirls.bc.edu/timss2023/frameworks/index.html

Murgatroyd, S. (2010). 'Wicked problems' and the work of the school. *European Journal of Education, 45*(2), 259–279. https://doi.org/10.1111/j.1465-3435.2010.01428.x

OECD. (2018). *The future of education and skills: Education 2030* (OECD Education Policy Perspectives, No. 98). OECD. https://doi.org/10.1787/54ac7020-en.

Reynolds, K. A., & Komakhidze, M. (2022). *TIMSS 2023 environmental attitudes and behaviors framework*. TIMSS & PIRLS International Study Center, Boston College. https://timssandpirls.bc.edu/timss2023/frameworks/pdf/T23-Environmental-Attitudes-and-Behaviors-Framework.pdf

Sachs, J. D., Lafortune, G., & Fuller, G. (2024). *The SDGs and the UN Summit of the future. Sustainable development report 2024*. Dublin University Press. https://doi.org/10.25546/108572

Sala, A., Punie, Y., Garkov, V., & Cabrera Giraldez, M. (2020). *LifeComp: The European framework for personal, social and learning to learn key competence*. European Commission: Joint Research Centre. Publications Office of the European Union. https://doi.org/10.2760/922681

Schleicher, A. (2021, January 25). *Green at fifteen—What schools can do to support the climate*. OECD Education and Skills Today. https://oecdedutoday.com/green-at-fifteen-schools-support-climate/.

Schulz, W., Ainley, J., Fraillon, J., Losito, B., Agrusti, G., Damiani, V., & Friedman, T. (2025). *Education for citizenship in times of global challenge. IEA International Civic and Citizenship Education Study 2022 international report*. Springer. https://doi.org/10.1007/978-3-031-65603-3.

Sipos, Y., Battisti, B., & Grimm, K. (2008). Achieving transformative sustainability learning: Engaging head, hands and heart. *International Journal of Sustainability in Higher Education, 9*(1), 68–86. https://doi.org/10.1108/14676370810842193

SOS International. (2021). *Students, sustainability and education. Results from a survey of students in higher education around the world*. https://www.sos.earth/posts/2020-survey.

Sterling, S. (2001). *Sustainable education. Re-visioning learning and change* (Schumacher Briefing, No. 6). Green Books.

Sterling, S. (2010). Living in the earth: Towards an education for our times. *Journal of Education for Sustainable Development, 4*(2), 213–218. https://doi.org/10.1177/097340821000400208

UNESCO. (2021). *Learn for our planet: A global review of how environmental issues are integrated in education*. UNESCO. https://unesdoc.unesco.org/ark:/48223/pf0000377362

UNFCC. (2022, April 13). *What is the triple planetary crisis?* https://unfccc.int/news/what-is-the-triple-planetary-crisis.

United Nations. (1999). *Resolution 1244 (1999). Adopted by the Security Council at its 4011th meeting, on 10 June 1999*. United Nations Security Council. https://digitallibrary.un.org/record/274488?ln=en.

Vesterinen, M., & Ratinen, I. (2024). Sustainability competences in primary school education–a systematic literature review. *Environmental Education Research, 30*(1), 56–67. https://doi.org/10.1080/13504622.2023.2170984

Vuorikari, R., Kluzer, S., & Punie, Y. (2022). *DigComp 2.2: The digital competence framework for citizens*. European Commission: Joint Research Centre. Publications Office of the European Union. https://doi.org/10.2760/115376

Wiek, A., Bernstein, M. J., Foley, R. W., Cohen, M., Forrest, N., Kuzdas, C., Kay, B., & Keeler, L. W. (2015). Operationalising competencies in higher education for sustainable development. In M. Barth, G. Michelsen, M. Rieckmann, & I. Thomas (Eds.), *Routledge handbook of higher education for sustainable development* (pp. 241–260). Routledge.

Yin, L., & Foy, P. (2021). Constructing the TIMSS 2019 environmental awareness scales. In M. O. Martin, M. von Davier, & I. V. S. Mullis (Eds.), *Methods and procedures: TIMSS 2019 technical report* (pp. 18.1–18.30). TIMSS & PIRLS International Study Center, Boston College. https://timssandpirls.bc.edu/timss2019/methods/chapter-18.html

Guia Bianchi is a researcher at the Joint Research Centre in Seville of the European Commission. Her research interests include competences and skills, and innovation policies. Among her achievements, she is a lead author of GreenComp and Sustainability competences, and coordinated the ACTIONbook on transformative innovation policies. She has a PhD in sustainability management and a Fulbright scholarship for research in the circular economy.

Ulrike Pisiotis works at the European Commission's Directorate General for Education, Youth, Sport and Culture on evidence-based policy and evaluation. Having joined the Commission in 2009, she has held various positions and gained extensive experience in education and research policies and systems within the EU and beyond, including learning for sustainability. She is also one of the co-authors of GreenComp, the EU sustainability competence framework. Ulrike Pisiotis has a PhD in German literature and has previously worked in research, publishing and teaching.

Paulína Koršňáková is a Senior Research and Liaison Adviser to IEA, and facilitates research networking, stakeholder engagement, and collaboration that enhance capacity building and knowledge sharing to foster innovation and quality in education. She serves as a series editor for the IEA Research for Educators and IEA Teacher Snippets. She holds a master degree in Teacher Preparation Programme of Biology and Mathematics and a doctorate in general psychology.

Chapter 2
Sustainability, Environment and Competencies to Act: Road Ahead for Science Education Across Europe

Jelena Radišić and Erik Knain

2.1 Setting the Stage

From a scientific viewpoint, it is widely recognized that human activity is driving major environmental challenges, including climate change, loss of biodiversity, and unsustainable consumption patterns. These challenges are intertwined with systemic and economic structures within society, as well as geopolitical developments, each of which causes turmoil in its own right. Adding to this complexity is a rapidly diminishing window of opportunity to avert potentially catastrophic tipping points (Intergovernmental Panel on Climate Change [IPCC], 2018, 2023; Intergovernmental Science-Policy Platform on Biodiversity and Ecosystem Services [IPBES], 2019). Thus, societies are confronted with the necessity for rapid and deliberate transitions towards sustainable pathways. IPBES acknowledges that both small-scale and large-scale changes can contribute to transformative change. It emphasizes that such change can be facilitated through cultural narratives and by altering dominant social norms, facilitating transformative learning processes, co-creating new knowledge, and weaving together different knowledge systems, worldviews, and values that recognize human–nature interdependencies and ethics of care (IPBES, 2024, p. 8). The crises pose profound challenges to education in general and science education in particular (Hodson, 2003; Tasquier et al., 2022), involving not only conceptualizing which competencies are vital for society today, but also recognizing that these competencies must evolve as new sustainable pathways that emerge in a rapidly changing world. Therefore, new competencies must not only accommodate change but also promote it. Such understanding has progressed alongside efforts within policy and assessment to evaluate competencies relevant to sustainability,

J. Radišić (✉) · E. Knain
Department of Teacher Education and School Research, University of Oslo, Oslo, Norway
e-mail: jelena.radisic@ils.uio.no; erik.knain@ils.uio.no

D. Hastedt et al. (eds.), *Educating Pupils for Environmental Sustainability in Europe*, IEA Research for Education 18,
https://doi.org/10.1007/978-3-032-14725-7_2

particularly focusing on its environmental aspects and the skills required for social actions addressing sustainability challenges.

This chapter provides a concise overview of how understandings of and significant discussions on environmental sustainability have changed over time and explores how current ideas connect with science education, influencing views and expectations in this field, especially regarding transformative change. Additionally, the chapter examines the concept of environmental sustainability competence and the initiatives to measure it (such as GreenComp), while adapting these to educational contexts (such as the Environmental Sustainability Competence Toolbox). The chapter concludes with an analysis of the Trends in International Mathematics and Science Study (TIMSS) 2023 Environmental Attitudes and Behaviors Framework and the interplay among the three frameworks in shaping the landscape for sustainability assessment.

2.2 From Environmental Education to Education for Transformative Change

One of the crucial challenges faced by mankind was articulated by the United Nations Brundtland Commission, which stated: "Humanity has the ability to make development sustainable to ensure that it meets the needs of the present without compromising the ability of future generations to meet their own needs" (Brundtland, 1987, Chapter 2, para. 1). This statement already indicates that making development sustainable is a complex concept that interconnects environmental, social, and economic aspects (United Nations, 2015). The complexity of these challenges affects the very nature of the problems and their potential solutions, rendering them "wicked,"[1] as they are deeply embedded in social, economic, and political systems and are characterized by contradictions, uncertainties, and the absence of straightforward solutions (Rittel & Webber, 1973), which can result in conflicting interests (Boström et al., 2018). The idea of development being sustainable and multifaceted is clearly articulated in the 2030 Agenda for Sustainable Development (United Nations, 2015), which encompasses 17 distinct Sustainable Development Goals (SDGs). The conceptual development stemming from the Brundtland definition represented a significant historical expansion of previous notions of nature and environmental preservation that were foundational to the modern environmental movement (Carson, 1962). By incorporating social and economic dimensions into discussions and understanding on how to make development sustainable, the largely United Nations-driven education for sustainable development (ESD) movement has

[1] A wicked problem involves multiple stakeholders with sometimes conflicting views, lacks a clear or mutually agreed problem statement, has no confirmed solutions, and may suffer from incomplete or missing knowledge. The problem can be entangled with assessments of risk, values, and ethics, and efforts to resolve it may alter the problem itself, leading to the emergence of new issues (Murgatroyd, 2010).

placed relatively more emphasis on localities and people. In contrast, environmental education (EE) focuses on the protection and enhancement of the environment (Pavlova, 2013). The inclusion of economic aspects in development has raised concerns about reducing poverty through economic development, leading to critiques of the ESD agenda for promoting unsustainable economic growth and being part of a neoliberal educational framework (Hickel, 2021; Jickling & Wals, 2008). A later concept, that of *sustainability* or *sustainability education*, can be viewed as an attempt to integrate both the EE and the ESD agendas, emphasizing the pursuit of a sustainable world without committing to specific conceptualizations of the processes to achieve it.

Recently, the concept of *transformative change* has gained prominence within the United Nations, particularly through the IPCC. The IPCC focuses on climate change mitigation and adaptation, while the IPBES emphasizes preserving nature and ecosystems. The United Nations Environment Programme (UNEP) has echoed this sentiment, stating: "Only a system-wide transformation will achieve well-being for all within the Earth's capacity to support life, provide resources and absorb waste. This transformation will involve a fundamental change in the technological, economic and social organization of society, including world views, norms, values and governance" (UNEP, 2021, p. 15). These developments reflect a growing emphasis on and understanding of the intricate dynamics within the areas outlined in the Brundtland definition. They effectively represent the concept of the Anthropocene. Humans have emerged as a planetary force reshaping the major life-supporting systems on a human timescale in the Anthropocene, contrasting with the geological timescale that characterizes the Holocene (Latour, 2017). This situation intertwines nature, worldviews, and economic paradigms.

With its emphasis on the interrelations between natural and societal systems and the need for deep-rooted change, the transformative change agenda can be viewed as an integration of both EE and ESD, sharing themes of lifelong learning, inclusion of social, environmental, and economic realms, collaboration between formal and informal learning, participatory learning, and higher-order thinking skills (Pavlova, 2013). The scope of the challenge it presents to education is profound. Transformative change education (TCE) challenges the curriculum to include competencies and perspectives that question and confront the more transmissive mode of teaching disciplinary knowledge, and ultimately, who and what education is for. TCE also presents a challenge for teachers who may be unfamiliar with teaching students future-oriented competencies, which poses a challenge for teacher education. The depth of the challenge extends beyond individual teachers; it encompasses the schools as organizations and professional communities. A common theme in educational policy documents advocates for schools to engage with the local community and collaborate with both formal and informal education providers, as well as involve students in real-life challenges.

The whole school approach (WSA) movement has emerged from an interest that extends beyond learning about sustainability confined to the classroom and embraces learning experiences through action in all aspects of school life (Mathie, 2024). The WSA is also notable for incorporating a multitude of experiences, perspectives, and skills in the best interests of students by engaging multiple stakeholders, including learners, parents, teachers, support staff, school leaders, community members, and service providers (United Nations Educational, Scientific and Cultural Organization [UNESCO], 2024a). The WSA significantly overlaps with *open schooling* as a strategy to foster collaboration between schools and out-of-school agencies. Open schooling refers to initiatives where schools collaborate with out-of-school entities such as enterprises, museums, universities, nongovernmental organizations, and municipalities to support students' inquiries into real-life sustainability challenges, aiming for community well-being (United Nations, 2015). Open schooling collaboration shows encouraging results in developing students' agency and their capacities for future thinking. A rare quantitative study in this area found a positive relationship between students' experiences in school of influencing society and their self-perceived action competence, which comprises knowledge of action possibilities, confidence in one's influence, and willingness to act (Torsdottir et al., 2024). Providing students with opportunities to connect scientific knowledge with resources for managing complexity and decision-making involving practical, personal, and political capabilities is also important for fostering their sense of agency (Tasquier et al., 2022); however, students have also been shown to express fragmented views about their abilities to effect change as actors in society (Røkenes et al., 2025).

The development described above, which emphasizes change and decision-making in real-life contexts, has been accompanied by efforts to identify sets of competencies regarded as generic across various settings, suitable for addressing issues that typically exist at the intersection of science and society, encompassing both political and ethical concerns. This approach demands a more comprehensive view of knowledge that does not separate scientific culture from humanistic culture (Hoffman, 2015). In the field of science education, the science, technology, and society movement (Aikenhead, 1994) was an influential push towards relating science education to societal and humanistic concerns (see the following section for an overview of different systems across Europe). Using a later conceptual framework by Sjöström and Eilks (2018), it resembles *Vision 3* of scientific literacy, which ultimately represents a science education with an overtly political dimension, "aiming at dialogical emancipation and socio-ecojustice, and emphasises transdisciplinarity, philosophical values, and praxis-oriented global citizenship" (Sjöström & Eilks, 2018, p. 77). Vision 3 builds upon the work of Roberts (2011), which was originally based on aims, or "emphases," regarding the purposes of science education in curriculum documents in Canada and the United States of America. Roberts later grouped one set of aims into *Vision 1*, focusing on scientific content and processes for later application, and another set of aims into *Vision 2*, addressing the applicability of scientific knowledge in everyday life and society (Roberts, 2011). While both visions are legitimate and can be found in nearly any broad school science curriculum, they compete for attention and may be in tension. Vision 3 is no different, with its emphasis on values, political aspects, and justice. Whereas Vision 2 challenges Vision 1 by emphasizing the everyday relevance and application of scientific knowledge and methods of inquiry, Vision 3 presents a more profound challenge by actualizing the deep economic and systemic fabrics of society, thereby bringing a political dimension to education. For instance, one implication of the concept of the Anthropocene is that the world can no longer be viewed as a passive container; in Hickel's (2021) words, as "a backdrop against which the human story plays out" (p. 78). Hickel argues that during the Enlightenment, the container view of the world led to the alienation of humans from the land and the objectification of nature, allowing ecosystems to become resources for an endless economic growth paradigm, which ultimately resulted in their degradation. Tapping into recurrent themes, particularly in the Nordic discourses on sustainability in education, degrowth is seen to necessitate reimagining the future role of schools towards an educational focus prioritizing sustainability, prudent resource use, and reshaping community values. This involves critical examination of ecological and economic systems, cultivating cooperative and communal competencies and knowledge, fostering values such as simplicity, conviviality, care, and a movement away from ideologies centered on the economic growth imperative (Berg-Brekkhus & Werler, 2024). Articulating the aims of Vision 2 and 3 of science education does not, by itself, provide answers to how to teach them. The development of students' scientific literacy and the importance of exploring and incorporating the societal dimension of science have been extensively examined in research studies within the field of socio-scientific issues (SSI) (see, for example, Kolstø, 2001; Sadler et al., 2017). These issues frequently involve risk and uncertainty as complicating factors (Sadler, 2009), and the relationship between science and society tends to be significant in teaching. Kolstø (2001) sought to extend beyond simple lists of essential skills for students in addressing SSI by discussing how various aspects of science are crucial to SSI. Kolstø identified eight content-transcendent topics, grouped into key areas: science as a social process, the limitations of science, values in science, and the critical attitude (see also Funtowicz & Ravetz, 1993).

SSI has been a vital area of focus in science education research, aimed at equipping young people with the civic scientific literacy essential for engaging in democratic dialogue and action on issues of public interest and decision-making (Herman et al., 2018; Zeidler et al., 2019). Efforts to synthesize decades of research on SSI have conceptualized socio-scientific reasoning as encompassing: (1) recognizing the inherent complexity of SSI; (2) examining issues from multiple perspectives; (3) appreciating that SSI are subject to ongoing inquiry; and (4) exhibiting skepticism when confronted with potentially biased information (Sadler et al., 2007). SSI serves as an effective context for enhancing students' motivation to learn science and for developing knowledge and processes that contribute to scientific literacy, including evidence-based argumentation, consensus building, moral reasoning, and the understanding and application of scientific knowledge (Evagorou & Nielsen, 2019). However, attributing a proactive meaning to the discussion of controversial issues may require students to possess the ability to solve complex problems by drawing on knowledge, values, skills, and attitudes that enable not only positioning in dialogue but also participation in effective action. This means that the SSI agenda in science education has grappled with similar challenges to those posed within the EE, ESD, and transformative change frameworks, specifically regarding the relationships between knowledge, nature, society, and education that foster capacities for transformative change.

2.3 Role of the School in Advocating for Transformative Change

Education is increasingly acknowledged as a vital long-term strategy for promoting sustainability (UNESCO, 2024b). Recent decades have seen the rise and development of global educational initiatives aimed at this goal, including EE, ESD and TCE. Each of these initiatives works to either challenge or uphold existing authorities, whether within established institutional frameworks or in opposition to them. Crucially, they promote a type of education that goes beyond merely sharing knowledge about sustainability. They highlight the significance of experiential learning in outdoor and real-world settings, which fosters awareness and nurtures pro-sustainability attitudes. This approach seeks to motivate and enable students to engage in collective action to tackle sustainability challenges at both local and global levels. As a result, these types of initiatives offer a framework that fundamentally questions the traditional logic and practices inherent in school curricula, prompting a reassessment of educational structures and delivery methods. Recognizing the need for radical transformation in how economies and societies function to adapt to and mitigate climate change, UNESCO (2024b) calls for a shift in educational paradigms from predominantly knowledge transfer to also include social, emotional, and action-oriented learning. Action-based, experiential, and collaborative learning is seen as key to motivating and enabling citizens to take climate action. Thus, the pressing nature of environmental issues has fostered a growing consensus on the critical need to prepare citizens for sustainable and environmentally friendly actions (Sass et al., 2023). Policymakers, researchers, and educators are increasingly discussing the competencies necessary to tackle environmental challenges and the methods to integrate and cultivate these within global educational systems. The complexities of the EE, ESD, and TCE frameworks particularly arise when assessing education systems' engagement with them.

International large-scale assessments (ILSAs), such as TIMSS and the Programme for International Student Assessment (PISA), provide extensive data on the knowledge domain of sustainability education (Reynolds & Komakhidze, 2022; White et al., 2023). In contrast, the International Civic and Citizenship Education Study (ICCS) offers valuable insights into attitudes and anticipated or intended behaviors related to sustainability (Damiani et al., 2025). Nevertheless, no single ILSA addresses all dimensions of any global EE, ESD, or TCE framework. This lack of cohesion presents a considerable challenge for the international community in its efforts to measure and monitor progress on this vital issue. Part of the challenge is the very notion of *competence*.

The Definition and Selection of Competencies (DeSeCo) project emerged as an interdisciplinary effort aimed at defining key generic functional competencies and outlining the nature of these competencies needed for individuals to act and achieve their objectives in specific contexts. DeSeCo aimed to inform (then) future ILSA projects. According to the DeSeCo project (Rychen & Salganik, 2003), competence (a) is the ability to master a complex challenge, (b) manifests in the actions of individuals in specific settings, and (c) is shaped by the characteristics of the task, challenge, or activity, for instance, the orchestration of attitudes, values, and skills that make effective social action possible.

The situated nature of competence is thus part of the challenge in measuring it. Individual competencies are context-dependent and situated, as they arise from the interaction between individuals and their physical and social environments. Therefore, competent action relies on the ability to adapt one's abilities to the specifics of the current situation while considering social motives and interests. Furthermore, any given concept of competence may be interpreted as either closely related to the situational context or more generic. The notion of *action competence* serves to illustrate this point. Action competence refers to an educational approach that is critical of moralistic tendencies, engages with democratic and participatory ideas in teaching, is problem-oriented and interdisciplinary, acknowledges potential conflicts of interest, and emphasizes lifestyle and living conditions (Jensen & Schnack, 2006; see also Sass et al., 2020). In DeSeCo, action competence is considered one of several key competencies. Mogensen and Schnack (2010) agree on certain points of DeSeCo, noting that competencies do not exist independently of context. However, it is still a category mistake to discuss action competence as being on the same level as other competences. Rather, it represents the sum of educational activities and goals that contribute to action competence as a broad educational vision at the level of Bildung.

A distinctive aspect of TIMSS is its connection to the science curricula of the participating education systems, addressing the complex nature of intended and enacted curricula. While the release of TIMSS data after each cycle attracts multiple empirical endeavors, countries also describe their systems alongside the extensive volume of data collected in each cycle. This includes a focus on subject curricula and other elements such as teacher professional development, compiled in the comprehensive TIMSS encyclopedia. The latest volume, published in 2024, provides an updated overview of whether sustainability, the environment, or agency is recognized in the national curricula of participating education systems across Europe, across the two grades TIMSS is administered, and how the intricacies of the EE, ESD, and TCE frameworks communicate with each other, if at all, within the intended curriculum (Reynolds et al., 2024).

Among the TIMSS 2023 participating education systems across Europe, based on the TIMSS encyclopedia records alone (Reynolds et al., 2024), almost all science curricula include at least the environmental aspect of sustainability (the exceptions are Albania, Bosnia and Herzegovina, Hungary and Kosovo) (Table 2.1). Moreover, the environmental aspect, in the context of preserving and safeguarding nature, is generally the most prominent aspect across the observed curricula. Simultaneously, the broader concept of sustainability is also widely present, although it may not always include all its aspects, such as preserving the environment, addressing social issues, making informed decisions, and shaping worldviews or visions for the future. In some cases, such as Norway, sustainability is also an interdisciplinary topic that connects various domains.

Table 2.1 Overview of the science curriculum in European education systems

	Grade four			Grade eight		
Education system	Sustainability (broadly)	Environmental focus	Transformative change	Sustainability (broadly)	Environmental focus	Transformative change
Albania						
Austria				•	•	
Belgium (Flemish)		•	•			
Belgium (French)	•	•				
Bosnia & Herzegovina						
Bulgaria		•				
Croatia						
Cyprus	•			•		
Czech Republic	•	•		•	•	
Denmark	•	•	•			
Estonia						
Finland	•	•	•	•	•	
France	•	•			•	•
Germany	•	•	•			
Greece						
Hungary				•	•	
Iceland						
Ireland	•	•		•	•	
Italy		•			•	
Kosovo						
Latvia	•	•				•
Lithuania	•			•	•	
Malta				•		
Moldova						
Montenegro		•				
Netherlands	•					
N. Macedonia, Rep. of		•				
Norway*	•			•		
Poland	•					
Portugal	•	•		•		
Romania		•		•		
Russian Federation						
Serbia		•				
Slovak Republic		•				
Slovenia		•				
Spain	•	•				
Sweden	•	•		•	•	
Switzerland						
Turkiye**		•			•	
Ukraine						

Notes: The shaded area indicates a European country that did not participate in TIMSS 2023 either with one or both target grades. Sustainability broadly refers to concepts related to preserving the environment, as well as social issues such as the future of communities and making informed decisions. It may emphasize worldviews and visions for the future. Environmental protection pertains to issues associated with safeguarding nature (e.g., common pollutants affecting the environment and the impact of human actions). Transformative change involves significant alterations, whether on a small or large scale, across various areas of society, along with the competencies needed to effect change, which might include values and worldviews

[a]Norway tests students in grades five and nine

[b]Turkiye assesses students in grades five and eight

Notably, a transformative framework involving the competencies needed to effect change, which might include values and worldviews, whether on a small or large scale, has penetrated school curricula (specifically, in Belgium [Flemish], Denmark, Finland, and Germany) (Table 2.1). However, this aspect still seems to pose a challenge for many countries. Its sparse visibility may very well reflect the varying levels of agency students possess across different education systems (Uitto et al., 2015). Monitoring and evaluating education that emphasizes transformative change and social action is, however, a significant challenge and must go beyond a simple focus on educational achievement to include values, ideologies, worldviews and political orientation as important for capability for action (UNESCO, 2024a).

Conversely, even if transformative change is intended in the curriculum, its enactment may still hinder its true nature. As policy has moved towards empowering students to take purposeful action (European Commission, Directorate-General for Education, Youth, Sport and Culture, 2022; UNESCO, 2020, 2024a), teaching that promotes action-oriented activities (Sass et al., 2023; Sinakou et al., 2019) will logically lead to the development of action competence (Sass et al., 2020) and the enactment of the transformative framework. However, if the teacher chooses which real-world problems the students will work on without a joint endeavor with the students, it will hinder the essence of the process and the intentions of the intended curriculum. Ultimately, this contributes to students potentially having fragmented views about their ability to effect change as societal actors (Røkenes et al., 2025).

2.4 Measuring Sustainability Competence

The integration of environmental, economic, and social aspects of sustainable development into school education has gained international significance, bolstered by the initiatives of various global organizations (European Commission, Directorate-General for Education, Youth, Sport and Culture, 2022; UNESCO, 2020). A prime example is UNESCO, which has actively promoted sustainability in educational systems. Initiatives such as the UN Decade of Education for Sustainable Development (2005–2014) (Buckler & Creech, 2014) and the ongoing Education 2030 Agenda (UNESCO, 2016) are instrumental in encouraging educational systems worldwide to embed ESD as a core component at all levels of education and in fostering competencies that enable young people to take action for sustainability.

In Europe, this emphasis is reflected in the Strategic Framework for European Cooperation in Education and Training towards the European Education Area (2021–2030) (Council of the European Union, 2021) and the European Commission's 2022 proposal on learning for environmental sustainability (European Commission, Directorate-General for Education, Youth, Sport and Culture, 2022). Moreover, the European sustainability competence framework, GreenComp, centers on competencies related to environmental sustainability (Bianchi et al., 2022). These initiatives have generated new momentum for implementing education concerning the environment and sustainability, marking a significant step in the evolution of educational policies towards a more sustainable future.

Despite a notable global rise in sustainable education (SE) policies and practices, it is evident that many education systems still do not incorporate SE in a systematic manner (see Sect. 2.3). Various studies have pointed out several challenges in different education systems, including the establishment of effective learning goals and suitable SE opportunities, the gathering of reliable data, and the creation of strong measurement tools to assess and monitor SE learning outcomes and processes (European Commission, Directorate-General for Education, Youth, Sport and Culture, 2022; Evans et al., 2012; Stepanek Lockhart, 2018; Taylor et al., 2019). Additionally, there is an acknowledged shortage of comparative data that would support empirical monitoring and evaluation of the effects of these educational strategies across nations, despite sustained efforts of ILSAs in contributing to these (Damiani et al., 2025; von Davier et al., 2024; White et al., 2023).

In the following section, three frameworks are examined, each attempting to address sustainability or sustainability with a focus on its environmental component. These include GreenComp (Bianchi et al., 2022), the Environmental Sustainability Competence Toolbox (Borgonovi et al., 2022a), and the TIMSS 2023 Environmental Attitudes and Behaviors Framework (Reynolds & Komakhidze, 2022). Notably, the frameworks discussed do not constitute an exhaustive list of all available frameworks, nor is it the intention to provide a comprehensive review of them, as this exceeds the aim of this chapter. Instead, three frameworks are presented, developed with different objectives, all situated within the education domain, even though not all are specifically designed for educational purposes. Ultimately, each framework represents a significant effort to capture and assess sustainability competence. They are explored in relation to EE, ESD, and TCE frameworks and visions of science, particularly Vision 2 and Vision 3.

2.4.1 GreenComp

Against the backdrop of the European Green Deal (European Commission, 2019), the European Skills Agenda for Sustainable Competitiveness, Social Fairness and Resilience (European Commission, 2020a), Achieving the European Education Area by 2025 (European Commission, 2020b) and the EU Biodiversity Strategy for 2030 (European Commission, Directorate-General for Environment, 2021), a European sustainability competence framework, GreenComp, was developed by the European Commission's Joint Research Centre team (Bianchi et al., 2022). The framework was conceived to empower individuals and communities to contribute to societal and environmental well-being.

GreenComp relies on the idea of sustainability, defining it as "prioritising the needs of all life forms and of the planet by ensuring that human activity does not exceed planetary boundaries" (Bianchi et al., 2022, p. 12). Similarly, sustainability competence was defined as a competence that "empowers learners to embody sustainability values, and embrace complex systems, in order to take or request action that restores and maintains ecosystem health and enhance justice, generating visions for sustainable futures" (p. 12).

GreenComp outlines 12 competencies organized into four interconnected areas: (1) *embodying sustainability values*, including the competencies *valuing sustainability*, *supporting fairness* and *promoting nature*; (2) *embracing complexity in sustainability*, including the competencies *systems thinking*, *critical thinking* and *problem framing*; (3) *envisioning sustainable futures*, including the competencies *futures literacy*, *adaptability* and *exploratory thinking*; and (4) *acting for sustainability*, including the competencies *political agency*, *collective action* and *individual initiative*. For a more comprehensive overview of the framework please see Chap. 1.

With its emphasis on values, complexity and action, GreenComp aligns with Vision 3 and SSI, and at the same time emphasizes EE's concern for ecosystems and all life forms, including but not limited to humans. This is a significant shift from the human-centered Brundtland definition. It also does not address *development*, which makes it different to both ESD and transformative change conceptualizations, although one only needs to compare the current situation and the sustainable condition set by the definition to recognize that development and transformative change are required (for example, for the area *acting for sustainability)*. Furthermore, GreenComp includes competencies like systems thinking, critical thinking, and problem-solving, along with skills in areas such as moral reflection. SSI has been regarded as a valuable entry point for sustainability education (Herman et al., 2018), providing opportunities to explicitly negotiate the multiple dimensions of complex environmental issues that involve ethical, political, and ideological considerations (for example, the unequal impacts of climate change).

However, frameworks like GreenComp point out potential shortcomings in the SSI approach, particularly the necessity to cultivate capabilities for not only thinking about but also taking action regarding these issues. Thus, the GreenComp framework highlights competencies that are less frequent in the SSI literature, such as political agency, collective action, and envisioning futures as distinct areas of competence.

2.4.2 Environmental Sustainability Competence Toolbox

Developed by the Organisation for Economic Co-operation and Development (OECD), the Environmental Sustainability Competence Toolbox serves as a practical framework aimed at providing individuals with the necessary knowledge, skills, attitudes, and behaviors to tackle environmental challenges and promote sustainable development (Borgonovi et al., 2022a).

Using the GreenComp framework as a basis, Borgonovi et al. (2022b) integrate it with the OECD's PISA framework and data to evaluate students' environmental sustainability competence and explore the factors associated with this competence.

Several factors must be considered to understand the OECD Toolbox. First, GreenComp articulates its definition of sustainability education through four competence areas, each encompassing a mix of knowledge, skills, and attitudes. Secondly, GreenComp broadly defines environmental sustainability competencies without delving into specific measurement issues. Notably, PISA began examining environmental awareness, attitudes, and behaviors in 2006 (OECD, 2006), prior to GreenComp's development. Consequently, PISA's operationalization of these aspects serves as a proxy for some competencies outlined in GreenComp. It should be recognized, though, that the knowledge dimension measured by PISA is not tuned towards sustainability competencies; rather, the "foundational levels in environmental sustainability competence" include that students achieve at least a proficiency level 2 in the PISA science test (OECD, 2023). Finally, it is important to note that attitudes and actual behavior are distinct yet interrelated concepts, which is why PISA also emphasizes measuring this connection (Borgonovi et al., 2022b).

Combining the GreenComp and PISA measures has resulted in the first GreenComp area, *embodying sustainability values*, being assessed through *caring about the environment* (Borgonovi et al., 2022b). The second area, *embracing complexity in sustainability*, is evident in *awareness of environmental issues* and *science proficiency*, while the third area, *envisioning sustainable futures*, is expressed through *environmental optimism* and *environmental self-efficacy*. Finally, the fourth area, *taking action for sustainability*, is demonstrated by *environmental behavior*. It is clear that a 1:1 correspondence was not established; however, given PISA's strong emphasis on students' application of scientific knowledge in life and society, aligned with Vision 2 and several elements represented in Vision 3, it represents an interesting attempt to operationalize and measure some elusive concepts. At the same time, aligning PISA measures with GreenComp reveals the subtle evolution of PISA measures, particularly between the 2015 and 2018 cycles. The former emphasizes the environmental focus in the measures, while the latter centers on global aspects, thus shifting more toward the ESD and TCE frameworks.

2.4.3 The TIMSS 2023 Environmental Attitudes and Behaviors Framework

The current climate crisis has profoundly shaped the development of ILSAs, shifting their focus to sustainability, particularly its environmental component (Reynolds & Komakhidze, 2022; White et al., 2023). As global discourse around environmental issues has intensified, so too has the need for educational assessments that tackle these critical topics. TIMSS 2019 addressed this growing urgency by retroactively adding an environmental scale, reflecting the heightened attention to environmental education in the international arena (Yin & Foy, 2020). With TIMSS 2023, the Environmental Attitudes and Behaviors Framework was introduced, which will be addressed in more detail given the overall organization and focus of this volume as well as the more comprehensive nature of the framework.

Traditionally, TIMSS has evaluated the environmental knowledge of students in fourth and eighth grades as part of its science assessment. For several cycles, the assessment has incorporated topics concerning the human impact on the environment and the use of natural resources. In TIMSS 2019, the evaluation of students' understanding of environmental issues was formalized with the introduction of the TIMSS 2019 environmental awareness scales (Yin & Foy, 2020). By identifying relevant items from the TIMSS 2019 science assessment that focus on environmental awareness, cognitive scales were constructed to gauge this awareness among fourth and eighth graders, strongly aligned with the EE framework.

Multiple studies demonstrate that addressing societal, political, and environmental issues necessitates a broader understanding of scientific facts (Olsson et al., 2022; Rieckmann, 2017; Wals, 2011). Acquiring knowledge does not inevitably result in practice changes; rather, such changes emerge when individual knowledge, thinking and attitudes are addressed (Glasman & Albarracín, 2006; Kollmuss & Agyeman, 2002; Macintyre et al., 2024; Olsson et al., 2022; Uitto et al., 2015; Vare & Scott, 2007). Classrooms and schools provide ideal environments for children to gain this knowledge (Olsson et al., 2022; Uitto et al., 2015). Moreover, teaching methods that encourage and nurture positive transformations in students' knowledge, attitudes, and behaviors are regarded as essential facilitators in this journey (Buckler & Creech, 2014). The TIMSS Environmental Attitudes and Behaviors Framework encompasses many of the above-mentioned elements, focusing on students, classrooms, schools, and the home environment.

Concerning students, the framework addresses both student attitudes and behaviors. The assessment of attitudes utilizes the two-dimensional Model of Ecological Values, which stems from the Theory of Ecological Attitude (Bogner & Wiseman, 1999, 2006; Wiseman & Bogner, 2003). This theory identifies two main dimensions of environmental attitudes: preservation and utilization. Preservation indicates a person's commitment to conserving and protecting nature, while utilization reflects the belief that natural resources exist for human exploitation, with faith in science and technology to resolve environmental challenges (Reynolds & Komakhidze, 2022). Students aligned with preservation typically enjoy nature and prioritize safeguarding natural areas. In contrast, those aligned with utilization often view nature as a resource for human gain and trust that scientific progress will address environmental issues. Assessment of attitudes shows a clear environment focus pertinent to EE and a focus on human needs, like the early Brundtland definition, unlike GreenComp. Preservation items are included in the student questionnaire for both fourth and eighth graders, whereas utilization is assessed solely for eighth graders.

Students' environmentally responsible behaviors include simple practices that individuals can adopt in various situations, like reusing items or alerting peers about actions that harm the environment. Thus, students are asked to report how often they perform these behaviors (Reynolds & Komakhidze, 2022). However, choices in developing behavior items were also shaped by the notion that children often have limited decision-making power and agency, with their ability to exert control differing across educational settings, cultures, and home backgrounds (Uitto et al., 2015). However, if unchallenged, this premise can hinder the development of action competence (Sass et al., 2020). A positive relationship between students' experiences in school of influencing society and their self-perceived action competence, which comprises knowledge of action possibilities, confidence in one's influence, and willingness to act (Torsdottir et al., 2024), indicate that the premise can be challenged by educational innovation, thus strengthening the TCE framework at school.

Regarding the school context, the TIMSS framework encompasses school policies and practices as well as classroom contexts that include teaching practices, attitudes, and competencies (Reynolds & Komakhidze, 2022). The framework acknowledges that a shared vision and a schoolwide approach to sustainability education can provide direction for teachers and promote collaboration in integrating sustainability into educational processes, a concept widely recognized in various studies (Breiting et al., 2005; Higgs & McMillan, 2006; Mathar, 2015; Tasquier et al., 2022). Additionally, such school practices are observed to significantly influence student actions and their development of action competence (Mathar, 2015; Olsson et al., 2022), thus supporting overall ESD goals. The TIMSS 2023 school questionnaire collects information on schools' emphasis on environmental sustainability for both fourth- and eighth-grade students. Principals are requested to indicate the extent to which their school participates in various activities that promote environmental sustainability (for example, reviewing school practices, encouraging relevant professional development for teachers).

The classroom context emphasizes teaching practices, attitudes, and competencies of teachers. Research indicates that students exposed to more active and interactive teaching methods—such as classroom discussions, research projects, hands-on activities, and outdoor environmental programs—show greater awareness of environmental issues, more positive attitudes toward the environment, and engage in environmentally responsible behaviors more frequently (Boeve-de Pauw et al., 2015; Røkenes et al., 2025; Uitto et al., 2015). Likewise, studies reveal that teachers' motivation and recognition of education's role in advancing sustainability significantly impact the effectiveness of sustainability education (Bertschy et al., 2013; Espinet et al., 2015; Varela-Losada et al., 2021). In line with this perspective, the TIMSS 2023 teacher questionnaire (Reynolds & Komakhidze, 2022) collects data on the techniques science teachers employ to educate students about sustainability and the degree to which they promote environmentally responsible behavior. Regarding teachers' attitudes and competencies, they express agreement on the priority of environmental and sustainability education in schools, alongside involvement in and the necessity for professional development to incorporate these topics into their teaching. Data is gathered from science teachers of both fourth and eighth grades.

Finally, the home context emphasizes parental practices (Reynolds & Komakhidze, 2022), highlighting that parents are key socialization agents for their children. Consequently, they significantly influence the development of their children's environmental attitudes (Leppänen et al., 2012). Simultaneously, studies indicate that parents' environmental concerns and behaviors affect their children's environmental attitudes and actions (Grønhøj & Thøgersen, 2009; Meeusen, 2014). TIMSS 2023 collects data on these direct socialization pathways through the home questionnaire. Parents of fourth-grade students are asked to report how frequently they participate in various activities with their child.

Overall, when considering the Environmental Attitudes and Behaviors Framework and the tension between EE, ESD, and TCE frameworks, current operationalizations strongly align with the EE and ESD elements; however, TCE is much less evident, particularly from the perspective of students. Considering the curricular element that dominates the TIMSS study and findings from the brief analyses of the science curricula in participating countries, the dominance of EE and ESD is not surprising. At the same time, the question arises as to what kind of operationalizations are needed in the TIMSS framework that would allow capturing TCE more comprehensively.

2.4.4 Linking the Frameworks

The comparison of the three frameworks highlights their common focus on environmental attitudes, knowledge, and pro-environmental actions, though their prioritization of these elements varies. Here, their alignment with SDGs, relevance to formal education, and assessment is also examined (Table 2.2).

GreenComp distinctly emphasizes systems/futures thinking and civic engagement, such as futures literacy and political agency, alongside values-driven competencies like fairness and care for nature (Bianchi et al., 2022). The OECD Toolbox places a unique focus on the emotional aspect, addressing youth's feelings and aspirations regarding the environment, as well as essential twenty-first century skills, including collaboration and digital/financial literacy (Borgonovi et al., 2022a). TIMSS organizes its framework according to the perspectives of survey respondents—students, parents, teachers, and principals—and utilizes a two-factor preservation/utilization attitude model (Reynolds & Komakhidze, 2022), which is not reflected in the other frameworks.

Overlaps exist in the emphasis on concern for the environment and the planet, reflected in GreenComp's nature promotion, TIMSS's preservation attitudes, and OECD's pro-environmental values. Additionally, they share a focus on grasping complex cause-and-effect relationships, evident in GreenComp's systems thinking and OECD's cognitive scale. While both GreenComp and OECD regard competences as comprehensive combinations of knowledge, skills, and attitudes, TIMSS emphasizes attitudes and behaviors alongside distinct contextual factors.

GreenComp notably creates a clear connection to SDG target 4.7 (ESD), while also addressing SDG 12 (responsible consumption) and SDGs 14–16 (promoting nature and justice and strong institutions). Although the OECD Toolbox does not explicitly connect to the SDGs, it highlights SDG 4 (quality education) and SDG 13 (climate action) as significant. Finally, TIMSS is driven by global objectives, with the introduction to its framework referencing the SDGs and the Paris Agreement as justification. The questionnaire items correspond with SDG 13 and SDG 4.7. Additionally, by evaluating school initiatives and teaching practices, it reinforces SDG 4 and SDG 17 (global partnerships in education).

Table 2.2 Comparison of the three frameworks

Aspect	GreenComp	OECD toolbox	TIMSS 2023 environmental framework
Alignment with SDGs	– Explicitly references SDG 4.7 – Implicitly relates to SDGs 12, 14, 15 and 16	– Not formally linked to SDGs – Implicitly relates to SDGs 4 and 13	– Directly motivated by global goals – Survey items align with SDG 13 and 4.7 – Implicitly relates to SDGs 4 and 17
Emphasis on EE, ESD and TCE	– Very strong ESD focus: explicitly includes social justice (fairness) and care for nature in its competences – Aims for transformative learning—*acting for sustainability* (political agency, collective action) and futures literacy encourage systemic change	– Primary focus is on EE – Highlights pro-environmental attitudes, knowledge and behaviors – Transformative aspects are limited; does not explicitly address the role of education in transformation	– Strong EE/ESD orientation – Highlights the potential of participatory instruction to promote change – Transformative change is implied (whole school and teacher capacity items)
Assessment	– No specific assessment tool is built in; GreenComp is a reference framework	– Utilizes the existing PISA framework and questionnaires to infer competencies – Emphasizes how questions on the PISA student questionnaire serve as specified dimension indicators	– Entire framework is a measurement tool – Questionnaire items (for students, parents, teachers, principals) to measure attitudes and behaviors
Relevance to formal education	– Designed for all levels and settings ("any learning context") – Can inform classroom objectives and cross-curricular projects on sustainability	– Focused with regards to upper-secondary education (around age 15) as it utilizes PISA data for 15-year-olds	– Directly tied to grades four and eight – Developed for formal school assessments – Indicators cover classroom teaching (science instruction), whole school practices, and home impact on school-age learners
Pedagogical approaches	– Competence-based, learner-centered approach – Remains non-prescriptive	– Does not propose instructional methods – Adopts a developmental, child-centered lens to interpret PISA data – Highlights transferable skills (e.g., collaboration) that align with the twenty-first century competency framework	– Includes items on active learning and teacher practices (e.g., inquiry, projects) but does not prescribe

Building on the earlier discussion of EE, ESD, and TCE, it is clear that GreenComp stands out as the most comprehensive framework for ESD and transformative change. It effectively weaves together environmental, social, and ethical dimensions, incorporating agency-oriented skills such as political agency, collective action, and adaptability—empowering learners to become change-makers (Bianchi et al., 2022). The OECD Toolbox (Borgonovi et al., 2022a) addresses several relevant aspects of ESD, including values, behaviors, and collaboration, but it is less comprehensive regarding systems thinking and fairness, focusing primarily on psychological and skill-based metrics. TIMSS excels in EE and aligns with ESD principles by considering factors involving schools and teachers, but it falls short in explicitly promoting transformative pedagogy; it evaluates current attitudes and actions without offering guidance on their development. Overall, GreenComp offers the most extensive range of sustainability competencies, particularly those associated with systemic change, while the Toolbox and TIMSS concentrate more on environmental attitudes and behaviors (Reynolds & Komakhidze, 2022).

While the analyzed frameworks were created for different aims, all are relevant to formal education. GreenComp mainly serves as a policy and reference framework, suitable for various levels and environments where learners are present. Furthermore, it can inform classroom objectives and interdisciplinary projects focused on sustainability. The Toolbox primarily targets upper secondary levels within education systems participating in PISA; however, the identified competencies are also applicable to lower secondary education as they represent outcomes from prior schooling. Similarly, TIMSS is associated with specific age groups, particularly grades four and eight. Some countries have implemented slightly different target grades (for example, Norway tests grades five and nine since the 2015 TIMSS cycle). The framework incorporates formal school assessments, including TIMSS context questionnaires, which feature indicators related to classroom teaching (science instruction), schoolwide practices, and home influences on school-age learners, thus making it clearly pertinent to formal education. Further, students' assessments on mathematics and science are clearly linked to the school curriculum, unlike PISA.

While none of the frameworks provide direct pedagogical guidance, common principles can be identified across all of them, even though they remain non-prescriptive. GreenComp emphasizes a competence-based and learner-centered approach. It advocates for interdisciplinary and systems approaches, such as fostering critical systems/futures thinking and civic engagement, but does not mandate specific methods. The OECD Toolbox relies on its analytical and data-driven nature, adopting student-centered lenses. While pedagogy remains open, it emphasizes the importance of education in developing environmental competence, highlighting transferable skills which align with the twenty-first century competency framework. In a manner akin to Toolbox, TIMSS also relies on its data, particularly regarding context. For instance, TIMSS 2023 featured items on active learning and teacher practices (such as inquiry and projects) but refrained from dictating specific pedagogies. Nonetheless, the accompanying rationale highlights established ESD methods: interactive, participatory instruction, modeling sustainable behaviors, and implementing whole school approaches.

2.5 Final Remarks

This chapter aims to offer a brief overview of critical discussions surrounding sustainability and the assessment of environmental sustainability. The chapter explored the concept of education for sustainability, highlighting the transition towards transformative change and how the evolving understanding of sustainability relates to its application within education, specifically in science. The next section delves into the assessment of sustainability, recognizing that it should be both comprehensive and informative for diverse stakeholders. This chapter does not intend to exhaustively address these topics, as they would necessitate an entire volume; instead, several essential points have been emphasized.

Firstly, shifting towards sustainable societies necessitates not only technical solutions but also a significant reorientation of social practices and livelihoods. Achieving this shift requires rethinking the objectives, content, competencies and purpose of education that would lead to such a society (Wals, 2011). Additionally, it is essential to cultivate understandings of sustainability that are part of capabilities for social action and can be integrated into various contexts that foster sustainable lifestyles, democratic engagement, active participation in socio-political matters, and human well-being (Cebrián et al., 2025; Hodson, 2003; Tasquier et al., 2022). The path briefly sketched, starting with Brundtland's (1987) conceptualization towards the idea of EE (Pavlova, 2013) and later ESD and sustainability education and/or TCE (Hickel, 2021; Tasquier et al., 2022), show how deeply engaged different actors were and still are in comprehensively addressing sustainability and how the topic penetrates education, policy, politics and everyday lives.

This brings forward the second point: the deeper the integration of ESD and transformative change into the formal educational system, and the greater the investment from governments and civil societies in this field, the more attention is directed towards assessing the effectiveness and efficiency of such contributions (Costa et al., 2024). Current endeavors through ILSA's offer a glimpse of how sustainability competence is showcased across the world; however, this glimpse is limited because not all countries and economies are part of such assessments. Even when they are, each assessment captures different aspects (for example, OECD's Toolbox and PISA constructs compared to TIMSS Environmental Attitudes and Behaviors Framework). Additionally, from the perspective of EE, ESD, and TCE, assessments still have difficulties capturing transformation and change. At the same time, assessment is ultimately a balancing act between the comprehensiveness of the conceptualization and the extent to which certain choices make assessments more complex and, thus, possibly less feasible.

Next, in discussing education as an arena for students to become equipped with competences that allows them to tackle real-life challenges for change-making, an important consideration is whether the idea of education is situated within the school walls alone or extended to opportunities for students to enact their action competence at the crossroads of formal, non-formal, and informal educational providers (United Nations, 2015). While it is known that the right knowledge and attitudes may foster an action to occur (Glasman & Albarracín, 2006; Kollmuss &

Agyeman, 2002; Macintyre et al., 2024; Olsson et al., 2022; Uitto et al., 2015; Vare & Scott, 2007), when schools engage students in community well-being initiatives while cooperating with local stakeholders, this fosters students' sense of agency (Tasquier et al., 2022) and confidence in their influence and willingness to act (Torsdottir et al., 2024).

Finally, this has significant implications for teachers and their daily practices in promoting sustainability, as teachers' sense of self-efficacy in undertaking this challenging task (Boeve-de Pauw et al., 2022) should hold equal importance to students demonstrating their action competence. Research indicates that students exposed to more active and interactive teaching methods develop a greater awareness of environmental issues and are more likely to engage in environmentally responsible behaviors (Boeve-de Pauw et al., 2015; Røkenes et al., 2025; Uitto et al., 2015). However, facilitating and orchestrating such activities requires teachers to possess a broader range of competencies, which necessitates adequate training not just for novice teachers. Therefore, when discussing the need to rethink the objectives, content, competencies, and purpose of science education (Wals, 2011), this rethinking must take place at all levels and should inclusively engage all participants in the process that shapes where education happens, how it unfolds, and what is assessed.

References

Aikenhead, G. S. (1994). What is STS science teaching? In J. Solomon & G. S. Aikenhead (Eds.), *STS Education international perspectives on reform*. Teacher's College Press.

Berg-Brekkhus, Å., & Werler, T. (2024). How can degrowth be a powerful driver for educational change—Seen from a Nordic point of view? *Nordic Studies in Education, 44*(4), 306–320. https://doi.org/10.23865/nse.v44.6849

Bertschy, F., Künzli, C., & Lehmann, M. (2013). Teachers' competencies for the implementation of educational offers in the field of education for sustainable development. *Sustainability, 5*(12), 5067–5080. https://doi.org/10.3390/su5125067

Bianchi, G., Pisiotis, U., & Cabrera Giraldez, M. (2022). *GreenComp: The European sustainability competence framework*. European Commission: Joint Research Centre. Publications Office of the European Union. https://doi.org/10.2760/13286

Boeve-de Pauw, J. B., Gericke, N., Olsson, D., & Berglund, T. (2015). The effectiveness of education for sustainable development. *Sustainability, 7*(11), 15693–15717. https://doi.org/10.3390/su71115693

Boeve-de Pauw, J. B., Olsson, D., Berglund, T., & Gericke, N. (2022). Teachers' ESD self-efficacy and practices: A longitudinal study on the impact of teacher professional development. *Environmental Education Research, 28*(6), 867–885. https://doi.org/10.1080/13504622.2022.2042206

Bogner, F. X., & Wiseman, M. (1999). Toward measuring adolescent environmental perception. *European Psychologist, 4*(3), 139–151. https://doi.org/10.1027/1016-9040.4.3.139

Bogner, F. X., & Wiseman, M. (2006). Adolescents' attitudes towards nature and environment: Quantifying the 2-MEV model. *Environmentalist, 26*, 247–254. https://doi.org/10.1007/s10669-006-8660-9

Borgonovi, F., Brussino, O., Seitz, H., Bertoletti, A., Biagi, F., Bitat, A., Karpinski, Z., & Montanari, M. (2022a). *The environmental sustainability competence toolbox: From leaving a better planet for our children to leaving better children for our planet* (OECD Social, Employment and Migration Working Papers, No. 275). OECD. https://doi.org/10.1787/27991ec0-en.

Borgonovi, F., Brussino, O., Seitz, H., Bertoletti, A., Biagi, F., Bitat, A., Karpinski, Z., & Montanari, M. (2022b). *Young people's environmental sustainability competence: Emotional, cognitive, behavioural, and attitudinal dimensions in EU and OECD countries* (OECD Social, Employment and Migration Working Papers, No. 274). OECD. https://doi.org/10.1787/1097a78c-en.

Boström, M., Andersson, E., Berg, M., Gustafsson, K., Gustavsson, E., Hysing, E., Lidskog, R., Löfmarck, E., Ojala, M., Olsson, J., Singleton, B. E., Svenberg, S., Uggla, Y., & Öhman, J. (2018). Conditions for transformative learning for sustainable development: A theoretical review and approach. *Sustainability, 10*(12), 4479. https://doi.org/10.3390/su10124479

Breiting, S., Mayer, M., & Mogensen, F. (2005). *Quality criteria for ESD-schools*. Austrian Federal Ministry of Education, Science and Culture.

Brundtland, G. H. (1987). *Report of the world commission on environment and development: Our common future*. United Nations General Assembly document A/42/427.

Buckler, C., & Creech, H. (2014). *Shaping the future we want. UN Decade of Education for Sustainable Development (2005–2014). Final report*. UNESCO. https://unesdoc.unesco.org/ark:/48223/pf0000230171.

Carson, R. (1962). *Silent spring*. Penguin.

Cebrián, G., Corres Gallardo, A. d. C., Renta Davids, A. I., & Junyent, M. (2025). Assessment of sustainability competencies in primary education: A systematic review. *The Journal of Environmental Education, 56*(3), 208–228. https://doi.org/10.1080/00958964.2025.2471967.

Costa, J., Alscher, P., & Thums, K. (2024). Global competences and education for sustainable development. A bibliometric analysis to situate the OECD global competences in the scientific discourse. *Z Erziehungswiss*. https://doi.org/10.1007/s11618-024-01220-z

Council of the European Union. (2021). Council Resolution on a strategic framework for European cooperation in education and training towards the European Education Area and beyond (2021–2030). *Official Journal of the European Union,* C 66/1–21. https://eur-lex.europa.eu/legal-content/EN/TXT/?uri=CELEX:32021G0226(01).

Damiani, V., Losito, B., Agrusti, G., & Schulz, W. (Eds). (2025). *Young citizens' views and engagement in a changing Europe. IEA International Civic and Citizenship Education Study 2022 European report*. Springer. https://doi.org/10.1007/978-3-031-68631-3.

Espinet, M., Junyent, M., Amat, A., & Castelltort, A. (2015). Moving schools towards ESD in Catalonia, Spain: The tensions of a change. In R. Jucker & R. Mathar (Eds.), *Schooling for sustainable development in Europe: Concepts, policies and educational experiences at the end of the UN Decade of Education for Sustainable Development* (pp. 177–199). Springer. https://doi.org/10.1007/978-3-319-09549-3_11

European Commission. (2019). The European Green Deal. COM(2019) 640 final. https://eur-lex.europa.eu/legal-content/EN/TXT/?uri=celex:52019DC0640.

European Commission. (2020a). European Skills Agenda for sustainable competitiveness, social fairness and resilience. COM(2020) 274 final. https://eur-lex.europa.eu/legal-content/EN/TXT/?uri=celex%3A52020DC0274&utm.

European Commission. (2020b). Communication from the Commission to the European Parliament, the Council, the European Economic and Social Committee and the Committee of the Regions on achieving the European Education Area by 2025. COM(2020) 625 final. https://eur-lex.europa.eu/legal-content/EN/TXT/?uri=celex%3A52020DC0625&utm.

European Commission, Directorate-General for Education, Youth, Sport and Culture. (2022). Proposal for a Council Recommendation on learning for environmental sustainability. COM/2022/11 final. https://eur-lex.europa.eu/legal-content/EN/TXT/?uri=CELEX:52022DC0011.

European Commission, Directorate-General for Environment. (2021). *EU biodiversity strategy for 2030: Bringing nature back into our lives*. Publications Office of the European Union. https://data.europa.eu/doi/10.2779/677548.

Evagorou, M., & Nielsen, J. A. (2019). *Socioscientific issues: Searching for new perspectives*. Paper presented at ESERA 2019, Bologna, Italy.

Evans, N., Whitehouse, H., & Gooch, M. (2012). Barriers, successes and enabling practices of education for sustainability in far North Queensland schools: A case study. *The Journal of Environmental Education, 43*(2), 121–138. https://doi.org/10.1080/00958964.2011.621995

Funtowicz, S. O., & Ravetz, J. R. (1993). Science for the post-normal age. *Futures, 25*(7), 739–755. https://doi.org/10.1016/0016-3287(93)90022-L

Glasman, L. R., & Albarracín, D. (2006). Forming attitudes that predict future behavior: A meta-analysis of the attitude-behavior relation. *Psychological Bulletin, 132*(5), 778–822. https://doi.org/10.1037/0033-2909.132.5.778

Grønhøj, A., & Thøgersen, J. (2009). Like father, like son? Intergenerational transmission of values, attitudes, and behaviours in the environmental domain. *Journal of Environmental Psychology, 29*(4), 414–421. https://doi.org/10.1016/j.jenvp.2009.05.002

Herman, B. C., Sadler, T. D., Zeidler, D. L., & Newton, M. H. (2018). A socioscientific issues approach to environmental education. In G. Reis & J. Scott (Eds.), *International perspectives on the theory and practice of environmental education: A reader, Environmental discourses in science education* (Vol. 3, pp. 145–161). Springer. https://doi.org/10.1007/978-3-319-67732-3_11

Hickel, J. (2021). *Less is more: How degrowth will save the world*. Windmill Books.

Higgs, A. L., & McMillan, V. M. (2006). Teaching through modeling: Four schools' experiences in sustainability education. *The Journal of Environmental Education, 38*(1), 39–53. https://doi.org/10.3200/JOEE.38.1.39-53

Hodson, D. (2003). Time for action: Science education for an alternative future. *International Journal of Science Education, 25*(6), 645–670. https://doi.org/10.1080/09500690305021

Hoffman, A. J. (2015). *How culture shapes the climate change debate*. Stanford University Press. https://doi.org/10.1515/9780804795050

IPBES. (2019). *Global assessment report on biodiversity and ecosystem services of the Intergovernmental Science-Policy Platform on Biodiversity and Ecosystem Services*. IPBES Secretariat. https://doi.org/10.5281/zenodo.6417333.

IPBES. (2024). *Thematic assessment report on the underlying causes of biodiversity loss and the determinants of transformative change and options for achieving the 2050 Vision for Biodiversity of the Intergovernmental Science-Policy Platform on Biodiversity and Ecosystem Services*. IPBES Secretariat. https://doi.org/10.5281/zenodo.17099472.

IPCC. (2018). *Global warming of 1.5°C. Special report*. https://www.ipcc.ch/sr15/.

IPCC. (2023). *Climate change 2023: Synthesis report*. Contribution of Working Groups I, II and III to the Sixth Assessment Report of the Intergovernmental Panel on Climate Change [Core Writing Team, H. Lee & J. Romero (Eds.)]. https://doi.org/10.59327/IPCC/AR6-9789291691647.

Jensen, B. B., & Schnack, K. (2006). The action competence approach in environmental education. *Environmental Education Research, 12*(3–4), 471–486. https://doi.org/10.1080/13504620600943053. Reprinted from *Environmental Education Research, 3*(2), 163–178 (1997).

Jickling, B., & Wals, A. E. J. (2008). Globalization and environmental education: looking beyond sustainable development. *Journal of Curriculum Studies, 40*(1), 1–21. https://doi.org/10.1080/00220270701684667

Kollmuss, A., & Agyeman, J. (2002). Mind the gap: Why do people act environmentally and what are the barriers to pro-environmental behavior? *Environmental Education Research, 8*(3), 239–260. https://doi.org/10.1080/13504620220145401

Kolstø, S. D. (2001). "To trust or not to trust, …"—pupils' ways of judging information encountered in a socio-scientific issue. *International Journal of Science Education, 23*(9), 877–901. https://doi.org/10.1080/09500690010016102

Latour, B. (2017). Anthropology at the time of the Anthropocene: A personal view of what is to be studied. In M. Brightman & J. Lewis (Eds.), *The anthropology of sustainability* (Palgrave studies in anthropology of sustainability) (pp. 35–49). Palgrave Macmillan. https://doi.org/10.1057/978-1-137-56636-2_2

Leppänen, J. M., Haahla, A. E., Lensu, A. M., & Kuitunen, M. T. (2012). Parent-child similarity in environmental attitudes: A pairwise comparison. *The Journal of Environmental Education, 43*(3), 162–176. https://doi.org/10.1080/00958964.2011.634449

Macintyre, T., Tilbury, D., & Wals, A. (2024). *Education and learning for sustainable futures: 50 years of learning for environment and change*. Taylor & Francis. https://doi.org/10.4324/9781003467007

Mathar, R. (2015). A whole school approach to sustainable development: Elements of education for sustainable development and students' competencies for sustainable development. In R. Jucker & R. Mathar (Eds.), *Schooling for sustainable development in Europe: Concepts, policies and educational experiences at the end of the UN Decade of Education for Sustainable Development* (pp. 15–30). Springer. https://doi.org/10.1007/978-3-319-09549-3_2

Mathie, R. G. (2024). A whole school approach: A synthesis of interconnected policy, practice, and research conceptualisations. In A. E. Wals, B. Bjønness, A. Sinnes, & I. Eikeland (Eds.), *Whole school approaches to sustainability* (pp. 9–33). Springer. https://doi.org/10.1007/978-3-031-56172-6_2

Meeusen, C. (2014). The intergenerational transmission of environmental concern: The influence of parents and communication patterns within the family. *The Journal of Environmental Education, 45*(2), 77–90. https://doi.org/10.1080/00958964.2013.846290

Mogensen, F., & Schnack, K. (2010). The action competence approach and the 'new' discourse of education for sustainable development, competence and quality criteria. *Environmental Education Research, 16*(1), 59–74. https://doi.org/10.1080/13504620903504032

Murgatroyd, S. (2010). 'Wicked problems' and the work of the school. *European Journal of Education, 45*(2), 259–279. https://doi.org/10.1111/j.1465-3435.2010.01428.x

OECD. (2006). *Assessing scientific, reading and mathematical literacy: A framework for PISA 2006*. OECD. https://doi.org/10.1787/9789264026407-en

OECD. (2023). *OECD Skills Outlook 2023: Skills for a resilient green and digital transition.* OECD. https://doi.org/10.1787/27452f29-en

Olsson, D., Gericke, N., & Boeve-de Pauw, J. (2022). The effectiveness of education for sustainable development revisited—A longitudinal study on secondary students' action competence for sustainability. *Environmental Education Research, 28*(3), 405–429. https://doi.org/10.1080/13504622.2022.2033170

Pavlova, M. (2013). Towards using transformative education as a benchmark for clarifying differences and similarities between environmental education and education for sustainable development. *Environmental Education Research, 19*(5), 656–672. https://doi.org/10.1080/13504622.2012.736476

Reynolds, K. A., Aldrich, C. E. A., Bookbinder, A., Gallo, A., von Davier, M., & Kennedy, A. (Eds.). (2024). *TIMSS 2023 encyclopedia: Education policy and curriculum in mathematics and science.* TIMSS & PIRLS International Study Center, Boston College. https://doi.org/10.6017/lse.tpisc.timss.rs5882.

Reynolds, K. A., & Komakhidze, M. (2022). *TIMSS 2023 environmental attitudes and behaviors framework.* TIMSS & PIRLS International Study Center, Boston College. https://timssandpirls.bc.edu/timss2023/frameworks/pdf/T23-Environmental-Attitudes-and-Behaviors-Framework.pdf.

Rieckmann, M. (2017). *Education for sustainable development goals: Learning objectives.* UNESCO. https://doi.org/10.54675/CGBA9153

Rittel, H. W. J., & Webber, M. M. (1973). Dilemmas in a general theory of planning. *Policy Sciences, 4*, 155–169. https://doi.org/10.1007/BF01405730

Roberts, D. A. (2011). Competing visions of scientific literacy. In C. Linder, L. Östman, D. A. Roberts, P.-O. Wickman, G. Erickson, & A. MacKinnon (Eds.), *Exploring the landscape of scientific literacy* (pp. 11–27). Routledge/Taylor & Francis. https://doi.org/10.4324/9780203843284

Røkenes, H., Jornet, A., & Knain, E. (2025). Young people envisioning desired futures through narratives of change in science education. *Futures, 169*, 103591. https://doi.org/10.1016/j.futures.2025.103591

Rychen, D. S., & Salganik, L. H. (Eds.). (2003). *Key competencies for a successful life and a well-functioning society*. Hogrefe & Huber.

Sadler, T. D. (2009). Situated learning in science education: Socio-scientific issues as contexts for practice. *Studies in Science Education, 45*(1), 1–42. https://doi.org/10.1080/03057260802681839

Sadler, T. D., Barab, S., & Scott, B. (2007). What do students gain by engaging in socioscientific inquiry? *Research in Science Education, 37*(4), 371–391. https://doi.org/10.1007/s11165-006-9030-9

Sadler, T. D., Foulk, J. A., & Friedrichsen, P. J. (2017). Evolution of a model for socio-scientific issue teaching and learning. *International Journal of Education in Mathematics, Science and Technology, 5*(2), 75–87. https://ijemst.com/index.php/ijemst/article/view/374

Sass, W., Boeve-de Pauw, J., Olsson, D., Gericke, N., De Maeyer, S., & Van Petegem, P. (2020). Redefining action competence: The case of sustainable development. *The Journal of Environmental Education, 51*(4), 292–305. https://doi.org/10.1080/00958964.2020.1765132

Sass, W., De Maeyer, S., Boeve-de Pauw, J., & Van Petegem, P. (2023). Effectiveness of education for sustainability: The importance of an action-oriented approach. *Environmental Education Research, 30*(4), 479–498. https://doi.org/10.1080/13504622.2023.2229543

Sinakou, E., Donche, V., Boeve-de Pauw, J., & Van Petegem, P. (2019). Designing powerful learning environments in education for sustainable development: A conceptual framework. *Sustainability, 11*(21), 5994. https://doi.org/10.3390/su11215994

Sjöström, J., & Eilks, I. (2018). Reconsidering different visions of scientific literacy and science education based on the concept of Bildung. In Y. J. Dori, Z. R. Mevarech, & D. R. Baker (Eds.), *Cognition, metacognition, and culture in STEM education: Learning, teaching and assessment* (pp. 65–88). Springer. https://doi.org/10.1007/978-3-319-66659-4_4

Stepanek Lockhart, A. (2018). Monitoring ESD: Lessons learned and ways forward. In A. Leicht, J. Heiss, & W. J. Byun (Eds.), *Issues and trends in education for sustainable development* (pp. 215–231). UNESCO. https://doi.org/10.54675/YELO2332

Tasquier, G., Knain, E., & Jornet, A. (2022). Scientific literacies for change making: Equipping the young to tackle current societal challenges. *Frontiers in Education, 7*. https://doi.org/10.3389/feduc.2022.689329

Taylor, N., Quinn, F., Jenkins, K., Miller-Brown, H., Rizk, N., Prodromou, T., Serow, P., & Taylor, S. (2019). Education for sustainability in the secondary sector—A review. *Journal of Education for Sustainable Development, 13*(1), 102–122. https://doi.org/10.1177/0973408219846675

Torsdottir, A. E., Olsson, D., Sinnes, A. T., & Wals, A. (2024). The relationship between student participation and students' self-perceived action competence for sustainability in a whole school approach. *Environmental Education Research, 30*(8), 1308–1326. https://doi.org/10.1080/13504622.2024.2326462

Uitto, A., Boeve-de Pauw, J., & Saloranta, S. (2015). Participatory school experiences as facilitators for adolescents' ecological behavior. *Journal of Environmental Psychology, 43*, 55–65. https://doi.org/10.1016/j.jenvp.2015.05.007

UNEP. (2021). *Making peace with nature: A scientific blueprint to tackle the climate, biodiversity and pollution emergencies*. https://www.unep.org/resources/making-peace-nature.

UNESCO. (2016). *Education 2030: Incheon declaration and framework for action for the implementation of sustainable development goal 4*. UNESCO. https://unesdoc.unesco.org/ark:/48223/pf0000245656

UNESCO. (2020). *Education for sustainable development: a roadmap*. UNESCO. https://unesdoc.unesco.org/ark:/48223/pf0000374802

UNESCO. (2024a). *How to include the whole school in foundational education for health and well-being*. Building strong foundations, Report 3. UNESCO. https://doi.org/10.54675/XFFX8745.

UNESCO. (2024b). *Education and climate change: Learning to act for people and planet*. UNESCO & Monitoring and Sustainability Education Research Institute, University of Saskatchewan. https://doi.org/10.54676/GVXA4765.

United Nations. (2015). *Transforming our world: The 2030 agenda for sustainable development* (A/RES/70/1). https://sdgs.un.org/2030agenda.

Vare, P., & Scott, W. (2007). Learning for a change: Exploring the relationship between education and sustainable development. *Journal of Education for Sustainable Development, 1*(2), 191–198. https://doi.org/10.1177/097340820700100209

Varela-Losada, M., Pérez-Rodríguez, U., Lorenzo-Rial, M., Vega-Marcote, P., & Reid, A. (2021). Dealing with global environmental change: The design and validation of the GEC attitude scale. *Environmental Education Research, 27*(1), 110–131. https://doi.org/10.1080/13504622.2020.1822990

von Davier, M., Kennedy, A., Reynolds, K., Fishbein, B., Khorramdel, L., Aldrich, C., Bookbinder, A., Bezirhan, U., & Yin, L. (2024). *TIMSS 2023 international results in mathematics and science*. TIMSS & PIRLS International Study Center, Boston College. https://doi.org/10.6017/lse.tpisc.timss.rs6460.

Wals, A. E. (2011). Learning our way to sustainability. *Journal of Education for Sustainable Development, 5*(2), 177–186. https://doi.org/10.1177/097340821100500208

White, P. J., Ardoin, N. M., Eames, C., & Monroe, M. C. (2023). *Agency in the Anthropocene: Supporting document to the PISA 2025 science framework* (OECD Education Working Papers, No. 297). OECD. https://doi.org/10.1787/8d3b6cfa-en.

Wiseman, M., & Bogner, F. X. (2003). A higher-order model of ecological values and its relationship to personality. *Personality and Individual Differences, 34*(5), 783–794. https://doi.org/10.1016/S0191-8869(02)00071-5

Yin, L., & Foy, P. (2020). Constructing the TIMSS 2019 environmental awareness scales. In M. O. Martin, M. von Davier, & I. V. S. Mullis (Eds.), *Methods and procedures: TIMSS 2019 technical report*. TIMSS & PIRLS International Study Center, Boston College. https://timssandpirls.bc.edu/timss2019/methods/chapter-18.html

Zeidler, D. L., Herman, B. C., & Sadler, T. D. (2019). New directions in socioscientific issues research. *Disciplinary and Interdisciplinary Scientific Education Research, 1*(11). https://doi.org/10.1186/s43031-019-0008-7

Jelena Radišić is a researcher at the Department of Teacher Education and School Research at the University of Oslo. In her work, she employs a variety of methods to observe processes related to learning and teaching, focusing on the interaction between different actors involved, namely students, teachers, and parents. She investigates topics such as motivation for learning, academic emotions, subject identity, and how specific instructional practices can support their development and overall learning. While her earlier work involved close collaboration with teachers on inquiry-based learning, action research, and reflexive practice, she has recently been involved in projects centred on open schooling collaboration between schools and out-of-school actors as a means to teach complex competencies that promote democratic participation, social action, and support future-oriented STEAM education for sustainability. Radišić is an associate editor for the European Journal of Psychology of Education, and from 2021 to 2025, she has been a Senior Coordinator for a special interest group of EARLI, SIG 10—Social Interaction in Learning and Instruction.

Erik Knain is a full professor of Science Education at the Department of Teacher Education and School Research (ILS), University of Oslo. He has long led projects involving collaboration between schools and researchers in inquiry-based teaching and learning, scientific literacy, socio-scientific issues, and sustainability education. He led the Horizon 2020 project "Science Education for Action and Engagement towards Sustainability" (SEAS) (2019–2022), which focused on open schooling collaboration between schools and out-of-school agents as a means to teach complex competencies that enable democratic participation and social action. He is currently a WP leader in the Erasmus+ project FEDORAS' Academy, which promotes future-oriented STEAM education for sustainability, coordinated by the University of Bologna. He has also worked on the concept of competence and was the Norwegian participant in the OECD DeSeCo project (Definition and Selection of Competencies). Commissions of trust include research leader at the Department of Teacher Education and School Research, Vice Dean of Research at the Faculty of Educational Sciences, University of Oslo, and he is presently an alternate member of the Board of Directors at the Centre for Advanced Study (CAS) in Oslo. He was also part of the working group for the Strategy for Climate and Environment at the University of Oslo.

Chapter 3
The European Sustainability Competence Framework (GreenComp) in Light of TIMSS 2023 Data

Paulína Koršňáková, Tahira Ali Qadri, and Wangqiong Ye

3.1 Introduction

Building upon the conceptual mapping of Trends in International Mathematics and Science Study (TIMSS) 2023 items to the GreenComp framework outlined in Chap. 1, this chapter adopts an empirical perspective to examine how students' reported behaviors and attitudes, as captured through the TIMSS 2023 context questionnaires, correspond to the sustainability competences defined in GreenComp. Briefly, the outcomes of the mapping in Chap. 1, which provide the foundation for the analysis conducted in this chapter, are presented and extended by exploring what the data reveal about the prevalence and distribution of sustainability-relevant dispositions among students across the European Education Area and partner education systems in the Western Balkans.

The overarching aim of this study is to assess the extent to which existing TIMSS questionnaire data can be interpreted in light of the GreenComp competence areas. In doing so, the study contributes to the broader objective of integrating sustainability considerations into the monitoring and evaluation of educational outcomes. More specifically, the study focuses on the attitudes, values, and behaviors of students that correspond to the four GreenComp competence areas: *embodying*

P. Koršňáková (✉)
International Association for Evaluation of Educational Achievements (IEA), Amsterdam, The Netherlands
e-mail: p.korsnakova@iea.nl

T. A. Qadri
IEA Amsterdam, Amsterdam, The Netherlands
e-mail: t.qadri@iea.nl

W. Ye
Center for Research on Equality in Education, University of Oslo, Oslo, Norway
e-mail: Wangqiong.ye@create.uio.no

D. Hastedt et al. (eds.), *Educating Pupils for Environmental Sustainability in Europe*, IEA Research for Education 18,
https://doi.org/10.1007/978-3-032-14725-7_3

sustainability values, *embracing complexity in sustainability*, *envisioning sustainable futures*, and *acting for sustainability* (Bianchi et al., 2022). By leveraging the TIMSS 2023 context questionnaire data, the analysis investigates whether and how these competences are reflected in students' self-reported experiences, perspectives, and learning environments. The focus on grade four students provides insight into early patterns of engagement with sustainability-related concepts, as well as the potential influence of home and school environments in shaping these orientations.

TIMSS is one of the longest-running international large-scale assessments, implemented by the International Association for the Evaluation of Educational Achievement (IEA). The study monitors trends in students' achievement in mathematics and science at the fourth and eighth grades, with assessments conducted regularly every four years since 1995. In addition to achievement tests that monitor international trends in mathematics and science, capturing the breadth of these subjects as they are taught in the participating education systems, TIMSS collects rich contextual data through extensive background questionnaires administered to national curriculum specialists and the school principals, students, teachers, and (for grade four students) parents. These questionnaires capture a wide range of information on the contextual factors at school and home which are associated with learning and students' attainment (Mullis et al., 2021).

From the perspective of sustainability education, TIMSS offers valuable opportunities to examine the conditions under which students acquire sustainability-related knowledge, develop environmentally friendly attitudes, and participate in ecologically responsible behaviors. For instance, the TIMSS 2023 context questionnaires included items on student enjoyment of nature, discussions about environmental issues at home and at school, instructional emphasis on conservation and biodiversity, and behavioral practices such as saving resources and reusing materials (Mullis et al., 2021). These data allow researchers to explore educational environments that support the development of sustainability competences and to examine cross-national patterns of engagement and equity in sustainability-related learning outcomes.

Although TIMSS is not explicitly designed to assess sustainability education, it is well positioned to contribute meaningfully to this field, particularly when interpreted through the lens of the GreenComp framework. GreenComp outlines a multidimensional vision of sustainability learning, built upon 12 interdependent competencies (Bianchi et al., 2022).

To explore the potential of TIMSS in capturing sustainability as a competence described by GreenComp, this study introduces the construct of *sustainability engagement* as its central analytical focus. Sustainability engagement is used to describe young (grade four) learners' interest and participation in, and attitudes toward, activities that support environmental stewardship and responsible interaction with natural systems. This construct is intentionally framed in alignment with the GreenComp framework. As operationalized in this study, sustainability engagement encompasses both affective dimensions (for example, enjoyment of nature, belief in protecting living systems) and behavioral components (for example, resource-saving practices, participation in environmental activities). It reflects a learner's emerging capacity to connect with nature, think critically about environmental issues, and participate actively in sustainability-related actions, elements broadly consistent with GreenComp areas such as *embodying sustainability values* and *acting for sustainability*, and specific competencies such as *promoting nature*.

The focus on sustainability engagement is also grounded in educational literature that emphasizes the importance of early behavioral and motivational pathways in the development of sustainability literacy and agency (Boeve-de Pauw et al., 2015; Olsson et al., 2016). Rather than focusing on knowledge acquisition or specific behavioral outcomes, as might be possible using TIMSS data, this more general, overarching concept follows the GreenComp narrative. In the present study, sustainability engagement is assessed through a selection of TIMSS 2023 questionnaire items that have been conceptually linked to GreenComp competences, enabling an empirical analysis of how early patterns of engagement vary across student populations and national contexts.

The study applies a structured methodology to measure and analyze sustainability engagement using TIMSS 2023 grade four student, home and teacher context questionnaire data. A subset of relevant items was selected based on conceptual alignment with GreenComp. Descriptive statistical analyses (means and standard deviations) provide an overview of item-level trends, followed by a latent class analysis (LCA) to identify distinct profiles of student sustainability engagement. These latent profiles are then examined across TIMSS 2023 participating education systems within the European Education Area and the Western Balkan region to explore cross-national similarities and differences in the manifestation of sustainability competences.

In addition to identifying student profiles, the chapter investigates the role of parents and teachers in shaping students' orientations toward sustainability. Drawing on additional items from the context questionnaires, the extent to which students perceive encouragement and support for sustainability-related activities at home and at school is explored. These findings are situated within broader discussions of how educational environments can foster the development of sustainability competences among learners.

The chapter is structured as follows:

- Section 3.2 reiterates the GreenComp framework and provides a brief overview of the analysis carried out in Chap. 1 and its results.
- Section 3.3 describes the data sources, outlines the criteria for item selection and recoding, and presents the analytical strategy, including the use of LCA, used in the present study.
- Section 3.4 presents the empirical findings, including descriptive statistics and latent class profiles, followed by an examination of their distribution across education systems and the role of contextual factors such as parental and teacher influence.
- Section 3.5 discusses the implications of the findings for educational monitoring, assessment design, and the integration of sustainability competences into large-scale international assessments.
- Section 3.6 concludes the chapter by summarizing key insights and suggesting directions for future research and policy development in the field of sustainability education.

3.2 Conceptual Frameworks Underpinning the Linking

3.2.1 Overview of the GreenComp Framework

The development of a European sustainability competence framework is one of the policy actions undertaken to promote learning on and for sustainability in the European Union. As education systems around the world increasingly seek to respond to urgent global challenges such as climate change, biodiversity loss, poverty, food insecurity, economic imbalances and social inequalities, there is growing recognition of the necessity to embed sustainability in education. The European sustainability competence framework—GreenComp—was developed by the European Commission to serve as a guiding reference for education and training systems in developing curricula, pedagogies, and learning outcomes that foster sustainability-oriented learning across lifelong education pathways in European Union Member States (Bianchi et al., 2022). It provides a non-prescriptive, adaptable structure that supports both formal and informal educational contexts, contributing to the European Green Deal's broader vision of a climate-neutral and environmentally sustainable Europe (European Commission, 2019). Designed to be inclusive of all learners regardless of age or educational sector, GreenComp provides a shared understanding of the knowledge, skills, and attitudes that constitute competence in sustainability (Bianchi et al., 2022).

Rather than prescribing curricular content or assessment models, GreenComp offers conceptual clarity and actionable guidance for educators, policymakers, and curriculum developers. Its premise is that education for sustainability requires a transformative shift that moves beyond knowledge acquisition to encompass the cultivation of values, behavioral dispositions, and a capacity for action. The framework's architecture reflects this integrated perspective, promoting lifelong learning goals that encourage learners to adopt sustainability as both a personal ethic and a societal responsibility (Bianchi et al., 2022).

3.2.1.1 Overview of the GreenComp Competence Areas

The GreenComp framework is organized into four interconnected competence areas, each comprising three specific competences (Table 3.1). These 12 competences are further elaborated and organized according to a knowledge–skills–attitudes structure, which aligns with widely accepted educational theory and frameworks (Bianchi et al., 2022).

These competencies are intentionally overlapping and interdependent, capturing the multidimensional and systemic nature of sustainability. As such, they are adaptable to various educational settings and cultural contexts (Bianchi et al., 2022).

While GreenComp is non-assessable in a standardized sense, it offers a conceptual frame for evaluating the extent to which learners are exposed to or demonstrate sustainability-related competences. Its structure supports curriculum development,

Table 3.1 GreenComp framework areas, competences and descriptors

Area	Competence	Descriptor
1. Embodying sustainability values *Focuses on personal values and ethical foundations of sustainability*	1.1 Valuing sustainability	Understanding and appreciating the significance of sustainable development
	1.2 Supporting fairness	Promoting social and intergenerational equity
	1.3 Promoting nature	Valuing biodiversity and ecological systems
2. Embracing complexity in sustainability *Develops the ability to perceive and analyze complex sustainability issues*	2.1 Systems thinking	Understanding interdependencies in natural and social systems
	2.2 Critical thinking	Evaluating claims and assumptions within sustainability debates
	2.3 Problem framing	Constructing and deconstructing sustainability problems from multiple perspectives
3. Envisioning sustainable futures *Fosters learners' ability to think creatively and anticipate future scenarios*	3.1 Futures literacy	Exploring plausible and preferred futures
	3.2 Adaptability	Adjusting one's thinking and behavior to changing conditions
	3.3 Exploratory	Imagining alternative solutions and pathways
4. Acting for sustainability *Encourages action-taking at individual and collective levels*	4.1 Political agency	Participating in democratic processes for sustainability
	4.2 Collective action	Collaborating with others for shared sustainability goals
	4.3 Individual initiative	Taking personal responsibility and action

Source: Based on *Table 1. GreenComp areas, competences, and descriptors* in Bianchi et al. (2022, pp. 14–15)

teacher education, and learning environment design, offering clear guidance for integration across subject areas, especially science education where themes of environmental awareness, resource use, and systems thinking are already embedded.

3.2.2 TIMSS Framework in Light of Sustainability Engagement

TIMSS offers a comprehensive framework for assessing educational outcomes and learning contexts in mathematics and science at both the grade four and grade eight levels. TIMSS is administered every 4 years and includes two major components: achievement assessments and contextual questionnaires. The achievement tests evaluate students' knowledge and skills (applying and reasoning) in mathematics and science, while the questionnaires collect information from students, teachers, school principals, and curriculum experts to provide a holistic view of the conditions influencing mathematics and science learning.

The TIMSS assessment framework in science (Mullis et al., 2021) includes content domains such as life science (biology), physical science (chemistry and physics), and Earth and environmental science, topics that naturally lend themselves to sustainability-oriented exploration. These align conceptually with elements of sustainability science and can be seen as precursors to sustainability competences as defined in GreenComp and similar frameworks (Boeve-de Pauw et al., 2015). In mathematics, domains such as data and chance, measurement, and geometry support the development of analytical, modeling, and problem-solving skills that are key to systems thinking and sustainability reasoning. The cognitive domains (knowing, applying, and reasoning) assessed across both subjects mirror competencies promoted in sustainability education, including evidence-based inquiry, critical thinking, and the ability to analyze complex interrelated systems.

Beyond cognitive testing, TIMSS is distinctive for its rich contextual information. The context questionnaires for students, teachers, and schools provide extensive data on learning environments, instructional practices, resource availability, curricular focus, student engagement, and home support systems (Mullis et al., 2021). For instance, at both grade levels, TIMSS gathers student self-reports on science interest, enjoyment of nature, resource-saving behaviors, and perceptions of school emphasis on environmental topics. Teachers are asked about their instructional focus, use of inquiry-based practices, and perceived curricular priorities. School leaders report on access to resources, emphasis on academic excellence, and school climate. Parents, in the home questionnaire (grade four only), provide information on sustainability-related behaviors at home and encouragement given to children to act responsibly toward nature (Mullis et al., 2021).

While TIMSS was not explicitly designed to measure sustainability education, many of its contextual items reflect these dimensions, such as valuing biodiversity, participating in nature-based learning activities, and experiencing pedagogical approaches that address real-world environmental issues. Moreover, TIMSS contextual data capture broader ecological and social conditions that influence learning, including parental support, teacher practices, and school-level emphasis on sustainability themes. As such, TIMSS provides a unique empirical foundation for examining how education systems foster the development of sustainability-related competencies among young learners.

The multi-layered structure of TIMSS provides numerous entry points for interpreting data in relation to the GreenComp framework. Both the mathematics and science assessments contain content and skills applicable to the development of sustainability competences. Likewise, of greater relevance to this chapter, the contextual questionnaires capture affective, behavioral, and environmental factors that influence the formation of sustainability-oriented attitudes and values. This chapter leverages this potential by interpreting selected TIMSS items through the GreenComp lens, in order to assess how sustainability-related values, behaviors, and learning conditions manifest among grade four students across diverse education systems. This approach aligns with emerging efforts to mainstream sustainability within assessment and monitoring systems (Brundiers et al., 2021; United Nations Educational, Scientific and Cultural Organization [UNESCO], 2021).

3.2.3 Observing Students' Sustainability Engagement Through TIMSS Questionnaire Data

TIMSS focuses on assessing students' achievement in mathematics and science at grades four and eight, while the GreenComp sustainability competence-related framework does not stipulate specific competences linked to science or mathematics education. Although certain science content areas, such as those related to ecosystems, energy, environmental changes, and human impact, are indirectly linked to sustainability, these concepts are embedded within a disciplinary science framework and are not explicitly developed from a sustainability-oriented perspective. While links between selected assessment items and areas 2 (*system thinking*, *critical thinking*, *problem framing*) and 3 (*adaptability* and *exploratory*) of the GreenComp framework could be established, the grade four core assessment offers limited opportunities to identify overlaps with sustainability-related concepts.

On the other hand, the TIMSS 2023 student context questionnaire provides several useful constructs and question clusters that can be leveraged as proxies to infer attitudes, beliefs, and behaviors related to nature, environment-friendly dispositions, and sustainable practices.

In relation to the grade four student questionnaire, it offers two questions, each with an extensive set of items, that stand out for their direct relevance to sustainability-related constructs (Table 3.2).

The first question (ASBG11A-E) probes students' emotional and affective connection to nature. Statements such as "I care about the protection of plants and animals," "It makes me sad when nature is destroyed" and "Addressing climate change should be a high priority" reveal students' empathy, concern for biodiversity, and early-level environmental consciousness and relate to attitudes toward nature and environment values.

The second question (ASBG12A-F) measures frequency of pro-environmental actions, including reuse and resource conservation, environmental discussions, participation in environmental group activities, and peer correction of unsustainable behavior. These items reflect both personal agency and developing social responsibility, corresponding to skills and behavioral competencies in sustainability. They allow researchers to analyze student engagement with sustainability as action, not just attitude. These correspond to self-reported sustainable practices of students at an early age.

Content available within the grade four student questionnaire focuses on developmentally appropriate constructs such as empathy, emotional connection to nature, and simple environmentally friendly actions. These questions reflect concrete, observable behaviors and attitudes that are meaningful and measurable at the primary education level.

Since grade four items offer broad cross-cultural relevance and interpretability, the questions in the grade four student questionnaire are ideally suited for constructing foundational indicators of sustainability engagement grounded in the affective and behavioral domains. By emphasizing universally accessible themes such as

Table 3.2 Questions within TIMSS 2023 grade four student questionnaire that help assess students' sustainability engagement

ASBG11A-E: How much do you agree with these statements?

Fill one circle for each line.

Statement	Agree a lot	Agree a little	Disagree a little	Disagree a lot
(a) I care about the protection of plants and animals	O	O	O	O
(b) It makes me sad when nature is destroyed	O	O	O	O
(c) I enjoy finding out what kinds of plants and animals live in my area	O	O	O	O
(d) I enjoy being in nature (e.g., forests, parks, deserts)	O	O	O	O
(e) Addressing climate change should be a high priority	O	O	O	O

ASBG12A-F: How often do you do these things to help the natural environment?

Fill one circle for each line.

Statement	Every day	Almost every day	Sometimes	Never
(a) I try to reuse things (e.g., <bags, bottles>)	O	O	O	O
(b) I try to use less resources (e.g., <water, food>)	O	O	O	O
(c) I talk about how to help the environment (e.g., <saving water, picking up trash>)	O	O	O	O
(d) I try to learn about environmental problems (e.g., <climate change, endangered animals>)	O	O	O	O
(e) I try to participate in group activities to help the environment	O	O	O	O
(f) I tell my friends when they are doing things that harm the environment	O	O	O	O

caring for nature and conserving resources, the questionnaire enables alignment with GreenComp across diverse education systems.

Therefore, the mapping exercise described in the next section, along with the subsequent analysis that forms the core of this chapter, draws on the grade four context questionnaires. These are considered most suitable for providing a clear and insightful exploration of students' sustainability engagement as reflected in the TIMSS 2023 data.

3.2.4 Mapping TIMSS 2023 Context Questionnaire Items to the GreenComp Framework

As covered in Chap. 1, a systematic item mapping process was undertaken to identify TIMSS 2023 grade four student, teacher and home questionnaire items that align with the GreenComp competence areas. The school questionnaire was excluded from this exercise due to its institutional, rather than personal, focus. The granular, item-level data available in the student, teacher, and home questionnaires provided a comprehensive basis for interpretation through the lens of the GreenComp framework.

The mapping involved thematic review of grade four questionnaire items to identify content aligned to the GreenComp areas, relevant to the concept of sustainability engagement. Then, each selected item was assigned to a GreenComp competence area using the framework's competence definitions to establish conceptual alignment. The individual mapping decisions were revisited and any discrepancies were reconciled through discussion.

Some selected items plausibly corresponded to more than one GreenComp area, while others revealed gaps in area representation (for example, overlaps between areas 4 and 1 or 4 and 2 while Area 3 was not represented in the home questionnaire). In addition, it was also challenging to pinpoint a specific competence within each area, where closer alignment would have been possible.

The resulting mapping revealed that a substantial number of items across all three questionnaires aligned with each of the GreenComp competence areas (Table 3.3; for a detailed overview of the mapped items, see Chap. 1, Sect. 1.5). The results of the mapping enable the presentation of complete GreenComp profiles for grade four students using TIMSS 2023 student questionnaire data only. This is possible because it remains at a higher level. However, this could not be repeated for parents (due to missing indicators for Area 3) or teachers (due to ambiguity in whether responses captured professional or personal attitudes and behaviors). This mapping forms the basis for the statistical analyses, including the development of latent class profiles described in the following sections.

This approach demonstrates how existing international assessment data can be repurposed to explore sustainability competences, an emerging priority in global education policy and research (Bianchi et al., 2022; United Nations Economic Commission for Europe, 2012).

3.3 The Present Study

This study employed LCA to explore patterns in students' sustainability engagement and examines variations across education systems. Furthermore, to better understand the underlying mechanisms of students' sustainability engagement, the study investigated the influence of both parental and teacher sustainability engagement. Specifically, it addressed the following research questions:

Table 3.3 Conceptual mapping of TIMSS 2023 questionnaire items to GreenComp competence areas

TIMSS 2023 source	GreenComp competence areas			
	Area 1. Embodying sustainability values	Area 2. Embracing complexity in sustainability	Area 3. Envisioning sustainable futures	Area 4. Acting for sustainability
Student questionnaire items				
ASBG11A	●			
ASBG11C	●			
ASBG11D	●			
ASBG12A				●
ASBG12B				●
ASBG12E				●
ASBG12F	●			
ASBS06			●	
ASBS07G		●		
ASBS08H		●		
Home questionnaire items				
ASBH09A		●		
ASBH09B[a]		●		●
ASBH09C[a]	●			●
ASBH09D	●			
ASBH09E[a]	●			●
Teacher questionnaire items				
ATBS02F[a]	●	●		
ATBS03A		●		
ATBS03B			●	
ATBS03E[a]		●	●	
ATBS04A	●			
ATBS04C[a]	●			●
ATBS05A	●			
ATBS05B[a]	●			●
ATBS05C[a]		●	●	
ATBS05D[a]	●	●		

Note: [a]Item was mapped to more than one competence area

1. How many distinct classes of sustainability engagement can be identified among students and what does it reveal about the level of their engagement?
2. Do students' sustainability engagement classes vary meaningfully by education system?
3. To what extent does parental and teacher environmental engagement predict students' class membership?

3.3.1 *Method*

3.3.1.1 Sample and Procedure

TIMSS 2023 assessed students corresponding to the fourth and eighth years of formal schooling in participating systems. The study employed a two-stage stratified cluster sampling design to achieve nationally representative samples. In the first stage, schools were randomly sampled within each participating education system, typically using probability proportional to size sampling based on enrollment data. In the second stage, intact classrooms within selected schools were sampled, and all students in those classrooms were invited to participate. The use of intact classrooms was particularly relevant for this study, as it allowed for the analysis of classroom-level influences on students' sustainability engagement alongside other factors.

The participation targets required by participating systems required at least 85% coverage of the national student population. While exclusions were permitted for students with severe disabilities, significant language barriers or for those attending schools in geographically inaccessible areas, the total exclusion rates were kept below 5% in most participating systems to ensure broad population coverage (von Davier et al., 2024).

This study utilized TIMSS 2023 grade four data from student, home, and teacher questionnaires. From these questionnaires, 25 items were chosen based on the mapping exercise as outlined in Sect. 3.2.4 and in Chap. 1. Of these, ten items focused on students' sustainability engagement, five addressed parental involvement in sustainable and environmental activities, and ten were connected to teachers' environmental engagement.

Next, the data for these 25 items were examined among European participants in TIMSS 2023 grade four, encompassing education systems from the Mediterranean and Balkans, through Western, Central and Nordic European regions. Due to the design of the questionnaire and administrative challenges, student items typically have less than a 15% missing rate among most European participants, whereas the missing rates for home and teacher items are significantly higher.

To optimize the utilization of the available data, it was structured into three distinct datasets (Table 3.4). The first dataset comprises 10 student questionnaire items collected from 25 European participants, each with a missing data rate of less than 15% for these items.

The second dataset is a subset of the first, including 13 European participants who also have less than 15% missing data on home questionnaire items. The third dataset is formed from a different selection of nine European participants from the first dataset, all with less than 15% missing data on teacher questionnaire items (Fig. 3.1).

Given the relatively high non-response rates in parental questionnaires in TIMSS, and in line with methodological recommendations that suggest missing data below 15% can be considered acceptable (see, for example, Schlomer et al., 2010; Tabachnick & Fidell, 2013), a 15% missingness threshold was applied to determine variable inclusion. The 15% missingness threshold was applied to these items to ensure data quality while maintaining sufficient sample sizes for the analysis. This threshold represents a widely accepted balance in educational research, minimizing

Table 3.4 Participants in three datasets

	Dataset 1	Dataset 2	Dataset 3
Student (*N*)	164,574	76,253	37,177
Female (%)	49.6	49.9	49.8
Descriptions	25 participants, 10 student items (<15% missing)	13 participants, 5 home items (<15% missing)	9 participants, 10 teacher items (<15% missing)
Participants			
Bulgaria	●	●	●
Cyprus	●	●	
Czech Republic	●	●	
Denmark	●		
Finland	●	●	●
Germany	●		
Hungary	●		
Ireland	●	●	●
Italy	●	●	●
Kosovo	●		
Latvia	●		
Lithuania	●		●
Montenegro	●		●
Netherlands	●		
Norway	●		
Poland	●	●	
Portugal	●	●	●
Romania	●	●	●
Serbia	●	●	
Slovak Republic	●	●	●
Slovenia	●	●	
Spain	●	●	
Sweden	●		
Belgium (Flemish)	●		
Belgium (French)	●		

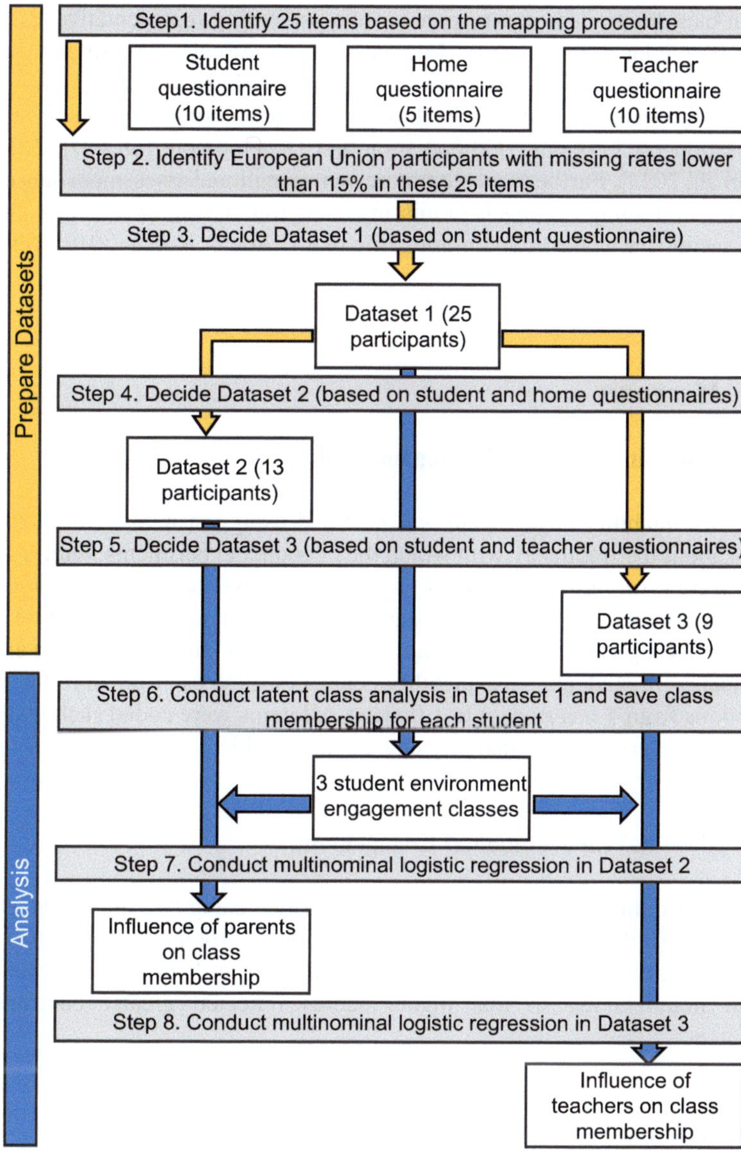

Fig. 3.1 A conceptual map presenting how datasets were prepared and used in analyses

potential bias due to item non-response while preserving representativeness across participating education systems. International large-scale assessments such as TIMSS and PISA routinely monitor item-level missing data and often consider thresholds between 10 and 20% during data processing and quality control (Organisation for Economic Co-operation and Development [OECD], 2017; von Davier et al., 2024). Furthermore, literature on international assessment data emphasizes that maintaining low levels of missing background data is critical for ensuring reliable secondary analyses and minimizing bias (Little & Rubin, 2019; Rutkowski & Rutkowski, 2010).

3.3.2 Measures

3.3.2.1 Student Sustainability Engagement

Student sustainability engagement was assessed using ten items. One item asked students about the frequency with which they conduct experiments, with responses ranging from 4 (*at least once a week*) to 1 (*never*). Five items required students to express their level of agreement with statements like "I enjoy being in nature," using a scale from 4 (*agree a lot*) to 1 (*disagree a lot*). The remaining four items explored the frequency of students' sustainability behaviors, such as "I try to reuse things," with options from 4 (*every day*) to 1 (*never*). All items were coded such that higher values indicate greater sustainability engagement (Table 3.5).

3.3.2.2 Parents' Environmental Engagement

Parents' environmental engagement was represented using five items that assessed the frequency of their interactions with children on environmental topics (for example, discussing the environment). Responses ranged from 1 (*never*) to 4 (*every day*), with all items coded so that higher scores reflected greater environmental engagement.

3.3.2.3 Teachers' Environmental Engagement

Teachers' environmental engagement was represented using ten items. One item assessed how frequently teachers asked students to engage in fieldwork, with responses ranging from 1 (*never*) to 4 (*every or almost every lesson*). Three items measured the extent to which teachers emphasized behaviors such as conducting experiments, using a scale from 1 (*not at all*) to 3 (*a lot*). Two items captured the frequency of teaching activities like developing positive attitudes, rated from 1 (*never or almost never*) to 4 (*at least once a week*). The remaining four items asked whether teachers addressed topics such as participating in environmental activities,

Table 3.5 Descriptive statistics

Dataset	Item ID	Item label	Mean	SD	Skew	Kurtosis
Dataset 1: Student items (10)	**ASBS06**	SCI\HOW OFTEN CONDUCT EXPERIMENTS	2.51	1.05	0.02	−1.19
	ASBS07G	SCI\AGREE\SCIENCE TEACHES ME	3.41	0.81	−1.35	1.19
	ASBS08H	SCI\AGREE\TEACHER ASKS TO SHOW	3.17	0.94	−0.88	−0.25
	ASBG11A	GEN\AGREE\PROTECTION OF PLANTS AND ANIMALS	3.63	0.66	−2.03	4.17
	ASBG11C	GEN\AGREE\LEARN ABOUT PLANTS AND ANIMALS	3.28	0.89	−1.06	0.21
	ASBG11D	GEN\AGREE\ENJOY BEING IN NATURE	3.59	0.7	−1.82	3
	ASBG12A	GEN\HELP ENVIRONMENT\REUSE THINGS	2.91	0.91	−0.24	−0.99
	ASBG12B	GEN\HELP ENVIRONMENT\SAVE RESOURCES	2.81	0.98	−0.26	−1.03
	ASBG12E	GEN\HELP ENVIRONMENT\ PARTICIPATE IN ACTIVITIES	2.43	1.08	0.21	−1.23
	ASBG12F	GEN\HELP ENVIRONMENT\TELL FRIENDS	2.73	1.06	−0.17	−1.26
Dataset 2: Home items (5)	**ASBH09A**	GEN\HOW OFTEN\DISCUSS ENVIRONMENT	2.29	0.65	0.94	0.97
	ASBH09B	GEN\HOW OFTEN\READ INFORMATION	2.11	0.62	0.86	1.88
	ASBH09C	GEN\HOW OFTEN\SAVE RESOURCES	2.88	0.81	−0.05	−0.9
	ASBH09D	GEN\HOW OFTEN\TIME IN NATURE	2.59	0.72	0.63	−0.61
	ASBH09E	GEN\HOW OFTEN\ENCOURAGE ACTION	2.74	0.82	0.17	−0.93
Dataset 3: Teacher items (10)	**ATBS02F**	SCI\ASK STUDENTS\DO FIELD WORK	2.39	0.75	0.69	0.06
	ATBS03A	SCI\EMPHASIS\ASK QUESTIONS	2.78	0.43	−1.56	1.06
	ATBS03B	SCI\EMPHASIS\PREDICT OUTCOMES	2.64	0.51	−0.89	−0.54
	ATBS03E	SCI\EMPHASIS\CONDUCT EXPERIMENTS	2.35	0.56	−0.09	−0.75
	ATBS04A	SCI\TEACHING\DEVELOP POSITIVE ATTITUDES	3.74	0.52	−1.96	3.21
	ATBS04C	SCI\TEACHING\DISCUSS ACTIONS	3.56	0.63	−1.13	0.24
	ATBS05A	SCI\TEACH ISSUES\VISIT NATURAL AREAS	1.76	0.43	−1.23	−0.49
	ATBS05B	SCI\TEACH ISSUES\PARTICIPATE ACTIVITIES	1.87	0.33	−2.26	3.09
	ATBS05C	SCI\TEACH ISSUES\RESEARCH TOPICS	1.76	0.43	−1.19	−0.57
	ATBS05D	SCI\TEACH ISSUES\OUTDOOR PROGRAMS	1.54	0.5	−0.14	−1.98

Note: *SD* standard deviation

coded as 1 (*no*) or 2 (*yes*). All items were coded so that higher values reflected greater environmental engagement.

3.3.3 Statistical Analysis

Data were prepared using R software version 4.0.2 (R Core Team, 2019). The *poLCA* package (Linzer & Lewis, 2011) was used to identify latent classes of students' sustainability engagement, while the *nnet* package (Ripley, 2022) was employed to examine the influence of parents' and teachers' environmental engagement on students' class membership. Missing data were addressed through listwise deletion, with both packages excluding cases with missing values in any of the variables included in the analysis (Lund & Ritz, 2025), resulting in models estimated on complete cases only. To account for the hierarchical structure of students nested within classrooms, a cluster bootstrap procedure was applied, treating classrooms as the clustering units when analyzing the influence of teachers (Lezhnina & Kismihók, 2022).

Specifically, the study first conducted an LCA on Dataset 1 (which included 25 participating education systems) to classify students' sustainability engagement using the *poLCA* package (Fig. 3.2). Comparisons across models of the following fit indices were used to decide the appropriate number of latent classes: the Akaike Information Criterion (AIC), the Bayesian Information Criterion (BIC), the sample-size adjusted BIC (SABIC), and the Lo–Mendell–Rubin likelihood ratio test (LMR). Parsimony and theoretical meaningfulness were also considered (Nylund et al., 2007).

This step aimed to uncover distinct patterns of students' sustainability engagement across education systems, allowing for a more nuanced understanding of how students cluster based on their behavioral and attitudinal responses. Identifying these latent classes provides a foundation for examining whether and how external factors—particularly parental and teacher engagement—are associated with

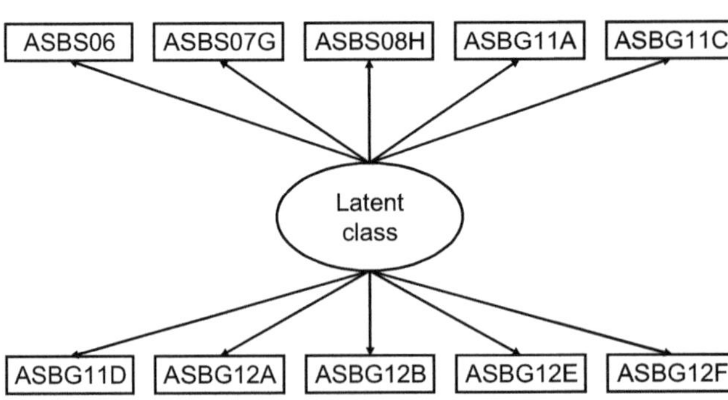

Fig. 3.2 A conceptual model of latent class analysis on student sustainability engagement

students' membership in different engagement profiles. In the next step, the extent to which parents' and teachers' ecological engagement predicted students' likelihood of belonging to each latent class was investigated.

The latent classes derived from this step served as the foundation for the subsequent phases. In the second step, the LCA results raised additional questions regarding the potential role of environmental engagement at home and in the classroom in shaping students' class membership. To address these questions, contextual data from parents and teachers were incorporated in the final phase. Specifically, parental and teacher environmental engagement variables were examined using multinomial logistic regression models to explore their association with students' class assignments.

Because the home and teacher questionnaire items had substantial missing data, and the education systems represented in Dataset 2 and 3 differed from those in Dataset 1, the LCA was conducted only once using Dataset 1. Each student was then assigned to their most likely latent class based on this initial analysis, and their class membership information was saved into Dataset 2 and 3 for subsequent analyses, avoiding the need to rerun LCA on datasets with more missing data or a different composition of participating education systems.

The influence of parents' environmental engagement on students' class membership, examined using multinomial logistic regression, was implemented with the *nnet* package on Dataset 2. Finally, the influence of teachers' environmental engagement on students' class membership also using multinomial logistic regression with an added cluster bootstrap procedure to account for the nesting of students within classrooms, was explored using Dataset 3.

3.4 Results

3.4.1 Classes of Students' Sustainability Engagement

3.4.1.1 Identifying Classes of Students' Sustainability Engagement

When performing LCA using Dataset 1 to identify emerging latent classes that describe the level of sustainability engagement among grade four students, distinct student profiles emerged. This enabled the assignment of each student included in the analysis to a respective class. However, before making these assignments, it was important to evaluate whether the identified classes were statistically robust and could be considered valid representations of the patterns observed in the data.

Based on the model fit indices (Table 3.4), the three-class solution was selected as the optimal model for identifying classes of students' sustainability engagement. Although the fit indices (AIC, BIC, and SABIC) continue to decrease with additional classes, the most substantial improvement occurs between the two-class and three-class models. This suggests that the three-class model offers a meaningful increase in model fit. Furthermore, while the two-class model has the highest

Table 3.6 Model fit indices for latent class analysis

Model	AIC	BIC	SABIC	Entropy	Smallest class %	LMR p-value
2-class	3,088,674.00	3,089,275.00	3,089,278.00	0.72	47.77	0.00
3-class	3,049,468.00	3,050,374.00	3,050,378.00	0.69	19.60	0.00
4-class	3,031,931.00	3,033,143.00	3,033,146.00	0.70	9.18	0.00
5-class	3,021,884.00	3,023,402.00	3,023,405.00	0.64	7.02	0.00
6-class	3,016,863.00	3,018,686.00	3,018,689.00	0.63	5.57	0.00

Notes: N = 164,574; *AIC* Akaike's Information Criterion; *BIC* Bayesian Information Criterion; *SABIC* Sample-Adjusted BIC; *LMR* Lo–Mendell–Rubin

entropy (0.72), the three-class solution still maintains a reasonably high entropy (0.69), indicating acceptable classification quality. Importantly, the three-class solution avoids the issue of very small class sizes seen in models with more than three classes—such as the four-class model, which includes a class with only 9.18% of the sample. Given the balance between improved model fit, classification precision, and meaningful class sizes, the three-class model was deemed the most parsimonious and interpretable solution (Table 3.6).

3.4.1.2 Description of the Three Classes of Students' Sustainability Engagement

Three latent classes were derived from students' responses to the ten selected items representing sustainability engagement of students (Fig. 3.3). As is common in LCA, the classes reflect distinct patterns of engagement, ranging from low to high levels of engagement with sustainability related behaviors, attitudes, and activities. Class 1, comprising the smallest proportion of students (19.6%), is marked by consistently low scores across all indicators of sustainability engagement taken into consideration. This group was labeled *Less Engaged* to reflect its limited involvement in environmental sustainability related activities. In contrast, Class 2, representing 35.65% of the sample, exhibited the highest scores across all ten items, indicating strong engagement; this group was labeled *Highly Engaged.* Finally, Class 3 included the largest proportion of students (44.75%) and displayed moderate scores across all items, suggesting a balanced pattern of involvement. Accordingly, this group was labeled *Moderately Engaged.*

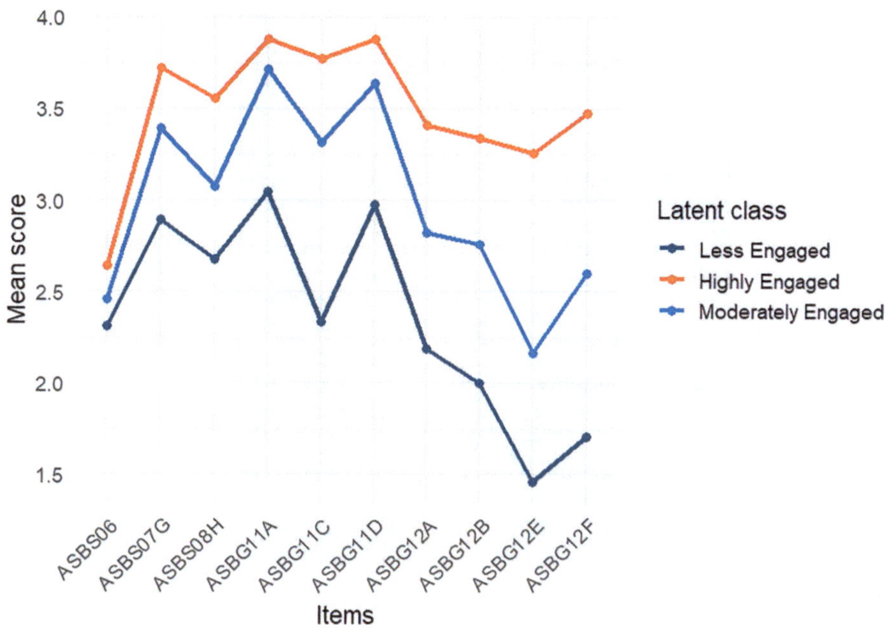

Fig. 3.3 Plots of three latent classes across ten items

3.4.2 Variation in Latent Classes by Education Systems and Gender

3.4.2.1 Class Membership Across Education Systems

A clear cross-system pattern emerged in the distribution of students across the three sustainability engagement classes (Fig. 3.4). Substantial variation was observed in the proportion of students assigned to each class, reflecting differing levels of sustainability engagement among students across education systems.

Class 1, representing students with low sustainability engagement, showed a significant degree of variation across education systems. Nordic education systems such as Denmark (40.11%), Norway (35.02%), and Sweden (34.27%) had the highest proportions of students in this least-engaged class. Similarly, many Western European education systems like Belgium (Flemish) (31.19%) and the Netherlands (39.56%) also had more than 30% of their students in this category. In contrast, several Eastern and Southern European education systems had much lower proportions of students in Class 1, such as Bulgaria (9.70%), Montenegro (8.28%), and Portugal (6.20%). Notably, Kosovo had the lowest percentage in Class 1, with only 5.15% of students falling into the Less Engaged profile. These patterns suggest a possible disconnect between traditionally high-performing or economically advanced systems and students' personal engagement with environmental issues.

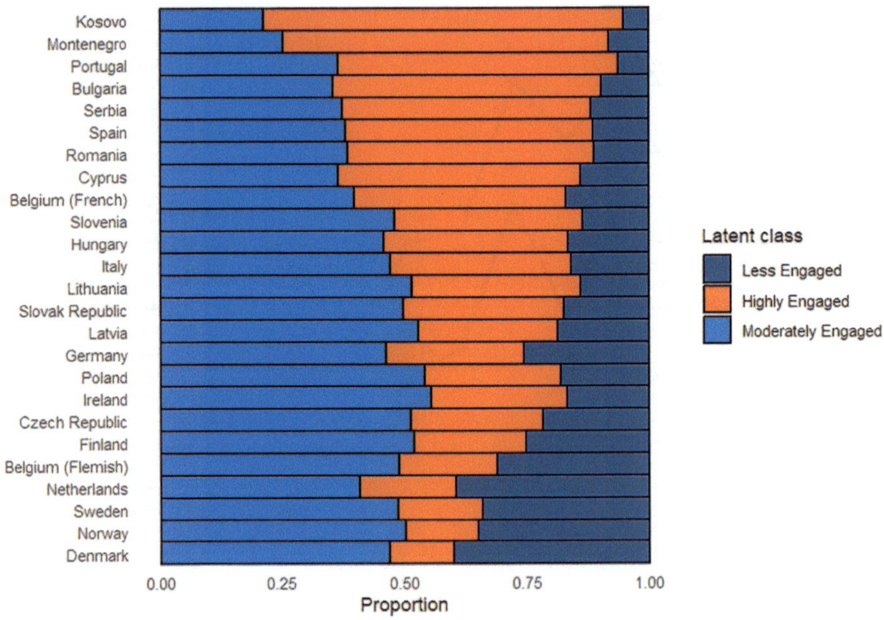

Fig. 3.4 Latent class distribution by education system. Note: Education systems are ranked from highest to lowest in Class 2 (Highly Engaged)

Class 2 represents the most ecologically engaged group and is particularly dominant in some of the newer or economically transitioning education systems (European Commission, 2008; OECD, 2024). The highest proportions of Class 2 students were found in Bulgaria (54.82%), Kosovo (73.29%), Montenegro (66.43%), and Portugal (57.18%). Additionally, Romania, Serbia and Spain each had over 50% of their student populations in Class 2. In contrast, education systems such as Belgium (Flemish) (19.91%), Denmark (12.87%), and Sweden (17.05%) showed notably lower proportions in this Highly Engaged group. These results may indicate a stronger focus on environmental issues in school curricula or community initiatives in some transitioning economies, or alternatively, differing interpretations of environmental engagement across cultural contexts.

Class 3, the Moderately Engaged profile, was the largest group in many Western and Northern European education systems. For example, in Finland (52.07%), Norway (50.20%), and Sweden (48.68%) the majority of students fell into this group. Similarly, the Czech Republic, Germany, and Italy also showed large shares of Moderately Engaged students. This pattern suggests that while students in these education systems may not exhibit the highest levels of sustainability engagement, a substantial number demonstrate at least a moderate commitment. Interestingly, in education systems with very high Class 2 proportions, such as Kosovo and Montenegro, the proportion of Class 3 students was notably lower (21.55% and 25.28%, respectively), indicating a clear contrast between high and low engagement levels in those contexts.

3.4.2.2 Class Membership Across Gender

Analysis of class membership by gender revealed a consistent pattern across education systems (Table 3.7): girls were disproportionately represented in the more engaged sustainability profiles—Class 2 (Highly Engaged) and Class 3 (Moderately Engaged)—while boys were overrepresented in Class 1 (Less Engaged). For example, in education systems like Finland and Sweden, over 54% of students in Class 2 were girls, and over 55% in Class 3. A similar trend was observed in education systems such as Lithuania, Norway, Poland, and Slovenia, where the percentage of girls in Class 2 ranged from approximately 51 to 54%, and in Class 3 from 53 to 56%. Conversely, boys constituted the majority in Class 1 across nearly all education systems—often exceeding 60%, such as in Germany (60.43%), Hungary (58.02%), and Kosovo (63.54%). These findings suggest that, across diverse national contexts, girls tend to show higher sustainability engagement than boys, with a consistent gender gap in favor of girls in the more engaged latent classes (Table 3.7).

Table 3.7 Class membership across gender

	Class 1		**Class 2**		**Class 3**	
Education system	**Girl (%)**	**Boy (%)**	**Girl (%)**	**Boy (%)**	**Girl (%)**	**Boy (%)**
Bulgaria	32.85	67.15	49.87	50.13	48.77	51.23
Cyprus	35.56	64.44	51.40	48.60	54.24	45.76
Czech Republic	34.47	65.53	51.64	48.36	54.67	45.33
Denmark	47.40	52.60	53.51	46.49	53.68	46.32
Finland	36.01	63.99	54.97	45.03	55.30	44.70
Germany	39.57	60.43	51.34	48.66	53.27	46.73
Hungary	41.98	58.02	50.66	49.34	52.45	47.55
Ireland	36.34	63.66	51.19	48.81	53.93	46.07
Italy	44.48	55.52	46.11	53.89	53.52	46.48
Kosovo	36.46	63.54	50.47	49.53	49.80	50.20
Latvia	38.98	61.02	49.78	50.22	55.85	44.15
Lithuania	35.53	64.47	53.33	46.67	53.40	46.60
Montenegro	37.04	62.96	50.07	49.93	48.67	51.33
Netherlands	47.23	52.77	49.51	50.49	57.52	42.48
Norway	43.00	57.00	50.98	49.02	56.63	43.37
Poland	35.87	64.13	53.74	46.26	53.93	46.07
Portugal	36.23	63.77	50.83	49.17	51.07	48.93
Romania	37.46	62.54	53.27	46.73	53.04	46.96
Serbia	36.53	63.47	52.70	47.30	51.81	48.19
Slovak Republic	42.37	57.63	49.71	50.29	51.35	48.65
Slovenia	35.74	64.26	48.98	51.02	55.68	44.32
Spain	39.30	60.70	51.45	48.55	49.56	50.44
Sweden	44.49	55.51	54.89	45.11	56.81	43.19
Belgium (Flemish)	44.58	55.42	47.04	52.96	52.00	48.00
Belgium (French)	42.08	57.92	50.94	49.06	55.01	44.99

Table 3.8 Parents' influence on students' class membership

		Class 2 vs. Class 1 (log-odds)	SE (Class 2)	p-value (Class 2)	Class 3 vs. Class 1 (log-odds)	SE (Class 3)	p-value (Class 3)
	Intercept	–1.934	0.059	< 0.001	–0.516	0.056	< 0.001
DISCUSS ENVIRONMENT	ASBH09A	0.156	0.024	< 0.001	0.15	0.023	< 0.001
READ INFORMATION	ASBH09B	0.289	0.025	< 0.001	0.117	0.024	< 0.001
SAVE RESOURCES	ASBH09C	0.142	0.016	< 0.001	0.092	0.016	< 0.001
TIME IN NATURE	ASBH09D	0.192	0.017	< 0.001	0.147	0.017	< 0.001
ENCOURAGE ACTION	ASBH09E	0.311	0.017	< 0.001	0.1	0.016	< 0.001

Note: *SE* standard error

3.4.3 Parents' Influence on Students' Class Membership

The results indicate that parental involvement plays a significant role in predicting students' sustainability engagement class membership (Table 3.8). Compared to students in Class 1 (Less Engaged), those in Class 2 (Highly Engaged) and Class 3 (Moderately Engaged) were more likely to have parents who frequently engaged in pro-environmental behaviors with them. All five parental variables—discussing environmental issues, reading information, saving resources, spending time in nature, and encouraging action—were positively and significantly associated with higher class membership ($p < 0.001$ across all comparisons).

Among these predictors, parental encouragement to take action showed the strongest association with Class 2 membership (log-odds = 0.311), followed by reading environmental information (log-odds = 0.289). Similar but slightly smaller effects were found for Class 3, suggesting that while these parental behaviors influence both moderate and high engagement, their impact is most pronounced for students in the Highly Engaged group. These findings underscore the important role of parents in shaping students' sustainability awareness and engagement through active involvement and environmental dialogue at home.

3.4.4 Teachers' Influence on Students' Class Membership

The multinomial logistic regression analysis examined the influence of science teaching practices on students' sustainability engagement class membership, comparing Class 2 (Highly Engaged) and Class 3 (Moderately Engaged) against Class 1 (Less Engaged) (Table 3.9). For Class 2, when teachers emphasized asking questions ($\beta = -0.09$, $p = 0.03$) and predicting outcomes ($\beta = -0.06$, $p = 0.04$), students were slightly less likely to be highly engaged. In contrast, practices involving discussing actions ($\beta = 0.27$, $p = 0.00$), visiting natural areas ($\beta = 0.14$, $p = 0.00$),

Table 3.9 Teachers' influence on students' class membership

		Class 2 vs. Class 1 (log-odds)	SE (Class 2)	p-value (Class 2)	Class 3 vs. Class 1 (log-odds)	SE (Class 3)	p-value (Class 3)
	Intercept	−1.91	0.28	0.00	0.04	0.04	0.28
ASK STUDENTS\DO FIELD WORK	ATBS02F	0.33	0.21	0.12	0.06	0.05	0.21
EMPHASIS\ASK QUESTIONS	ATBS03A	−0.09	0.04	0.03	0.01	0.04	0.83
EMPHASIS\PREDICT OUTCOMES	ATBS03B	−0.06	0.03	0.04	−0.12	0.07	0.07
EMPHASIS\CONDUCT EXPERIMENTS	ATBS03E	0.09	0.06	0.13	−0.02	0.05	0.75
TEACHING\DEVELOP POSITIVE ATTITUDES	ATBS04A	−0.06	0.05	0.23	0.11	0.08	0.20
TEACHING\DISCUSS ACTIONS	ATBS04C	0.27	0.06	0.00	0.01	0.07	0.84
TEACH ISSUES\VISIT NATURAL AREAS	ATBS05A	0.14	0.05	0.00	0.22	0.06	0.00
TEACH ISSUES\PARTICIPATE ACTIVITIES	ATBS05B	0.15	0.05	0.00	0.07	0.05	0.12
TEACH ISSUES\RESEARCH TOPICS	ATBS05C	0.11	0.04	0.01	0.38	0.06	0.00
TEACH ISSUES\OUTDOOR PROGRAMS	ATBS05D	0.19	0.05	0.00	0.14	0.05	0.00

Note: *SE* standard error

participating in environmental activities ($\beta = 0.15, p = 0.00$), teaching research topics ($\beta = 0.11, p = 0.01$), and organizing outdoor programs ($\beta = 0.19, p = 0.00$) were positively and significantly associated with high engagement. These results suggest that hands-on, applied learning experiences, such as outdoor and research-based activities, are more effective in fostering students' sustainability engagement than purely classroom-based questioning strategies.

When comparing Class 3 (Moderately Engaged) to Class 1 (Less Engaged), fewer teaching practices were significantly associated with higher sustainability engagement. The most notable predictor was teaching research topics ($\beta = 0.38$, $p = 0.00$), which showed a strong positive relationship with moderate engagement. In addition, visiting natural areas ($\beta = 0.22$, $p = 0.00$) and organizing outdoor programs ($\beta = 0.14$, $p = 0.00$) was also significantly related to higher odds of being in Class 3. Other teaching practices, such as asking questions, conducting experiments, or discussing environmental actions, did not significantly differentiate Class 3 from Class 1. These findings suggest that while more intensive hands-on or inquiry-based teaching strategies may not consistently distinguish Moderately Engaged students, exposure to research-related and outdoor learning experiences plays an important role in elevating environmental involvement above the lowest engagement group.

3.5 Discussion

This chapter examined the sustainability engagement of grade four students in 25 European Education Area and Western Balkan education systems using TIMSS 2023 data, interpreted through the lens of the GreenComp framework for sustainability competencies. Through LCA, cross-national comparisons, and regression modeling, the study identified three student profiles and the influence of home and school environments on their sustainability engagement. The findings reveal significant variation across the European Education Area and its Western Balkans neighborhood, demographic subgroups and in relation to familial and pedagogical contexts, offering important insights into the broader dynamics of sustainability education.

3.5.1 Profiles of Student Sustainability Engagement

The LCA identified three meaningful patterns of student engagement with sustainability topics and behaviors. The Highly Engaged group (35.7%) reflected students with strong pro-environmental attitudes and frequent sustainability behaviors, while the Moderately Engaged group (44.8%) demonstrated intermediate levels of engagement, suggesting some awareness and participation but without consistent depth. The Less Engaged group (19.6%) displayed low scores across both affective and behavioral dimensions of sustainability engagement.

These distinctions support a developmental view of sustainability learning, wherein attitudes and behaviors evolve in tandem with cognitive, emotional, and contextual inputs. Students in the Highly Engaged group may be experiencing early internalization of sustainability values, potentially reinforced by supportive home and school environments. The presence of a sizable Moderately Engaged group further indicates that many students occupy a transitional space, open to sustainability messages but perhaps lacking consistent reinforcement or opportunities for action.

Emerging research emphasizes that emotional connection to nature may play a key role in developing sustainability engagement (Otto & Pensini, 2017). Positive affective experiences, such as joy, fascination, and comfort in natural settings, have been shown to foster "significant life experiences" (Chawla, 1998) that promote enduring pro-environmental values and behaviors (Ernst & Theimer, 2011). As such, liking nature may often precede caring about nature, and ultimately motivate action. These emotional experiences may serve as an essential foundation upon which cognitive understanding and behavioral competencies can build. Students in the Highly Engaged group may thus reflect early internalization of sustainability values, likely reinforced not only by knowledge and social support, but also by positive emotional experiences with nature. This underscores the importance of ensuring that sustainability education integrates direct experiences with nature as a means of fostering emotional agency, and long-term commitment to sustainability action (Ardoin et al., 2019).

3.5.2 Cross-National and Demographic Patterns

3.5.2.1 Variation Across Education Systems

Substantial variation emerged in the distribution of engagement profiles across education systems. Interestingly, education systems typically associated with strong educational performance and higher levels of economic development, such as Denmark, the Netherlands, and Norway, had larger proportions of Less Engaged students. In contrast, higher levels of engagement were observed in Southern and Eastern European contexts, such as Bulgaria, Kosovo, Montenegro, and Portugal.

This pattern may reflect differences in how sustainability behaviors are embedded into daily life across contexts. The nature of sustainability learning and its interaction with socio-technological contexts must be explored. In some education systems, sustainability-related practices, such as energy conservation, water management or waste reduction, are increasingly integrated into the design of public infrastructure, technology, and service provision (such as automated lighting, centralized heating systems, efficient water regulation, or organized waste collection). These systemic solutions may reduce the frequency with which individuals, including students, are required to make active choices regarding resource use in their everyday routines (Shove, 2010; Stern, 2000). In other contexts, where such structural solutions are less common or widespread, sustainability behaviors may remain more directly tied to individual practices and family routines, creating more frequent opportunities for students to engage in sustainability-related actions and discussions (Boeve-de Pauw et al., 2015).

In addition, some highly developed systems may primarily integrate sustainability-related topics in science, geography or social studies curricula as knowledge domains, rather than as personal, behavioral or engagement-oriented instruction. This knowledge-oriented approach, emphasizing understanding over action, may develop cognitive understanding without necessarily translating into personal behavioral engagement, especially if students perceive sustainability as an overarching societal responsibility rather than a personal one (Olsson et al., 2016).

Moreover, in many transitioning economies, students may encounter more visible and immediate sustainability challenges in their environment necessitating swift action and a heightened awareness of and engagement with sustainability-related systems.

3.5.2.2 Gender Differences

Across almost all education systems, somewhat more girls than boys were found in the Highly Engaged and Moderately Engaged classes, while boys were somewhat more frequent in the Less Engaged group. This pattern aligns with longstanding findings in environmental education literature, which suggest that girls are more likely to internalize communal and care-oriented values, including environmental stewardship.

This trend reinforces the importance of gender-responsive sustainability education. Although the GreenComp framework does not explicitly address gender, the competencies it emphasizes—such as critical thinking, collaboration, and personal responsibility—may resonate differently with boys and girls depending on pedagogical framing and socialization. Addressing gender gaps in sustainability engagement may require not only inclusive curricular content but also differentiated instructional strategies that foster relevance and agency for all students. In addition, these patterns underscore the potential to channel girls' strong engagement with sustainability topics into greater participation in STEM (science, technology, engineering, and mathematics) education and professions, where such competencies are increasingly essential for addressing complex environmental and ecological challenges. At the same time, by intentionally aligning sustainability topics with STEM learning, education systems may be able to transform existing interest into long-term participation, helping also to close persistent gender gaps in STEM fields.

3.5.2.3 Parental Education and Home Contexts

While TIMSS did not directly measure parental education, its likely correlation with home-based sustainability practices was evidenced in the regression analysis in this study. Prior research has repeatedly demonstrated that parental education correlates strongly with student academic achievement, largely through mechanisms of cognitive stimulation, language exposure, and enriched learning environments (Mullis et al., 2020; Yang Hansen & Gustafsson, 2016). In the domain of sustainability education, however, these cognitive mechanisms may be complemented, and in some cases superseded, by value-based processes. Numerous studies have shown that higher parental education is associated with stronger pro-environmental attitudes, greater ecological concern, and more frequent sustainable behaviors at home (Franzen & Vogl, 2013; Lee et al., 2020; Zelezny et al., 2000). These value orientations likely shape not only what children know about sustainability, but also how they prioritize, internalize, and act on sustainability issues (Chawla & Cushing, 2007; Stevenson et al., 2013).

This dual role of parental education—as both a provider of knowledge and a transmitter of values—may be particularly relevant for sustainability engagement, which requires the integration of cognitive, affective, and behavioral competencies (Bianchi et al., 2022; UNESCO, 2020). While parallels with academic achievement exist, sustainability learning entails additional socio-emotional dimensions, including empathy, moral reasoning, and future-oriented thinking (Boeve-de Pauw et al., 2015; Olsson et al., 2016). Thus, the observed association between parental education and students' sustainability engagement underscores the importance of home-school partnerships that not only address knowledge gaps but also support the development of pro-environmental values and agency, particularly for students whose home environments may offer fewer such opportunities.

3.5.3 The Role of Parents in Shaping Engagement

Parental environmental engagement emerged as a significant predictor of students' likelihood to be in the more engaged classes. Specifically, parental behaviors such as discussing environmental topics, spending time in nature, saving resources, and encouraging children to take action were strongly associated with higher levels of student engagement. This finding aligns with well-established research in educational and environmental psychology demonstrating that parents serve as both role models and facilitators in the development of children's environmental values, behaviors, and knowledge (Chawla & Cushing, 2007; Evans et al., 2018; Grusec & Hastings, 2015).

These findings reinforce ecological socialization theory, which indicates that children's values and behaviors are shaped through daily interactions within their proximal environments (Grusec & Davidov, 2007). In the context of sustainability, parents influence children not only through direct instruction but also through modeling pro-environmental behaviors, communicating value-driven narratives, and providing experiential learning opportunities (Evans et al., 2018). Parental practices that combine behavioral modeling with dialogic engagement, such as conversations about environmental problems or encouraging pro-environmental actions, appear particularly influential in fostering children's emerging sense of sustainability agency, an essential competence embedded in the GreenComp domain *acting for sustainability*. Importantly, this study contributes to a growing body of research that positions parents as key co-educators in the development of children's sustainability competences. More broadly, such parental involvement may contribute to children's developing understanding of sustainability not only as a set of individual behaviors, but as an integrated concept involving environmental, social, and economic dimensions (Bianchi et al., 2022; United Nations, 2015). Through everyday practices and conversations, parents help children make sense of complex interconnections between resource use, collective responsibility, and long-term consequences, thereby laying an early foundation for systems thinking and future-oriented decision-making, both central pillars of sustainability competence frameworks (Boeve-de Pauw et al., 2015; Wiek et al., 2011).

These findings carry important implications for sustainability education practice and policy. While schools serve as central venues for formal learning, many aspects of sustainability competence may develop more effectively when learning is extended into the home environment through family-based activities, school–home partnerships, and community engagement initiatives (Barr, 2007; Evans et al., 2018; Wals, 2015). Educational policies aiming to broaden the reach of sustainability education should consider integrated approaches that actively involve families in sustainability learning, ensuring that both knowledge and values are nurtured from multiple sources of influence during students' formative years.

3.5.4 The Role of Teachers and Teaching Practices

The influence of teaching practices on environmental engagement was both significant and nuanced. The most impactful pedagogies were experiential and context-rich, including outdoor learning, participation in environmental programs, and project-based activities. These were positively associated with high levels of sustainability engagement, consistent with constructivist theories of learning (Piaget, 1972; Vygotsky, 1978), which emphasize that meaningful learning occurs when students actively engage with authentic, real-world contexts and construct personal understanding through interaction and reflection.

Experiential approaches provide opportunities for students to develop multiple dimensions of sustainability competencies, especially those covered in GreenComp, such as problem-solving, responsibility, and connection with nature. As previous research has shown, authentic learning environments foster students' motivation, agency, and self-efficacy in addressing complex sustainability challenges (Boeve-de Pauw et al., 2015; Evans et al., 2012). Moreover, outdoor and project-based pedagogies engage students both cognitively and affectively, facilitating deeper emotional connections to sustainability topics that serves as a prerequisite for behavioral change (Ardoin et al., 2019).

In contrast, the study found that more traditional classroom strategies, such as asking questions or predicting outcomes, had weak or negative associations with engagement. This suggests that pedagogical mode may matter as much as content: didactic approaches may not be sufficient to foster deep engagement with sustainability issues, particularly among younger learners who benefit from embodied and participatory experiences. Active, embodied, and participatory strategies appear better suited to cultivating the motivation and skills associated with sustainability.

Interestingly, several strategies showed stronger effects for students in the Highly Engaged class compared to the Moderately Engaged, hinting at a threshold effect:

once students are meaningfully engaged, certain pedagogical inputs may further amplify their involvement. However, raising students from low to moderate engagement may require broader structural and systemic interventions, such as fostering a supportive school climate, embedding sustainability throughout the curriculum, providing equitable access to nature experiences, and strengthening home–school–community partnerships (Evans et al., 2012; UNESCO, 2017).

Collectively, these findings underscore that teaching for sustainability is both pedagogically and systemically demanding. While active, experiential practices are demonstrably effective, their implementation often depends on teacher training, curriculum flexibility, institutional support, and cultural contexts that value student agency and participatory learning. As such, advancing sustainability education requires not only pedagogical innovation but also organizational commitment to fostering whole school approaches that create enabling environments for both teachers and students.

3.5.5 Linking Student Performance and Sustainability Engagement

This chapter, by taking a first step in identifying how sustainability engagement is represented in TIMSS questionnaires, offers a way forward in bridging comparative large-scale assessments and sustainability frameworks.

An important area for future research lies in examining how students' sustainability engagement profiles relate not only to their achievement in science subjects but also to their broader environmental literacy. While this chapter has focused on interpreting TIMSS 2023 questionnaire items through the lens of the GreenComp framework, the inclusion of achievement data could provide a more comprehensive understanding of the interaction of cognitive, affective, and behavioral dimensions of sustainability competence.

Further support for this perspective comes from the recent TIMSS Insights report, *Environmental Awareness in TIMSS 2023: Patterns in Achievement, Attitudes, Behaviors, and Contexts*, which presents cross-national analyses of students' knowledge about environmental issues, their attitudes and behaviors toward sustainability, and how these relate to science achievement, and home and school contexts (Bookbinder et al., 2025). The report highlights that student engagement in sustainability-related behaviors varies considerably across education systems and does not consistently align with science achievement patterns. For instance, some high-performing education systems in science report comparatively modest levels of environmental action among students, while others with lower academic performance show stronger behavioral engagement (Bookbinder et al., 2025). These findings suggest that while strong science achievement may provide a necessary foundation for sustainability engagement by contributing to knowledge and reasoning skills, it is not sufficient on its own. Environmental literacy, values-based education, and opportunities for experiential engagement also play an important role in fostering students' capacity and motivation to act sustainably. Future research could build on this insight by exploring how these three dimensions (achievement, environmental literacy, and behavioral dispositions) interact to shape students' preparedness for sustainability-oriented thinking and action.

3.6 Conclusion

This chapter explored how international large-scale assessment data, specifically the TIMSS 2023 grade four context questionnaires, can be leveraged to operationalize the GreenComp framework and assess students' sustainability engagement levels. Through a systematic item mapping and empirical analysis, the study identified how selected TIMSS 2023 student, home, and teacher questionnaire items reflect the core GreenComp competence areas, offering new insights into early patterns of sustainability-related dispositions among young learners across education systems in the European Education Area and Western Balkans.

At the core of this study is the construct of sustainability engagement, which was defined as students' reported interest and participation in, and attitudes toward, behaviors that support environmental sustainability. This includes both affective aspects (such as enjoyment of nature) and behavioral aspects (such as resource-saving practices and participation in environmental activities), aligned conceptually with GreenComp domains.

Using LCA, the study identified three distinct profiles of student sustainability engagement: Less Engaged, Moderately Engaged, and Highly Engaged. These profiles illustrate the multifaceted and developmental nature of sustainability competences, integrating affective, behavioral, and contextual dimensions that evolve through students' interactions with their educational and home environments. The cross-national variation observed in the distribution of these engagement profiles underscores how sustainability competences are shaped by diverse curricular, cultural, and socioeconomic factors.

Notable gender patterns emerged across participating education systems, with girls exhibiting relatively higher sustainability engagement than boys. These results call for approaches in sustainability education that foster inclusivity, motivation, and agency for all learners. Such approaches may also serve as a pathway for enhancing girls' interest and participation in STEM fields, where sustainability-related competences are increasingly relevant.

The findings further highlight the influential role of parents and teachers in shaping students' sustainability engagement. Parental behaviors such as discussing environmental issues, demonstrating pro-environmental practices, and encouraging children to take sustainability actions were strong predictors of higher engagement levels. Similarly, pedagogical approaches that integrate experiential, outdoor, and project-based learning were positively associated with students' engagement, emphasizing the value of authentic learning contexts that foster emotional connection to nature, agency, and active participation.

While the study provides important empirical contributions, several aspects warrant careful consideration. First, the analysis was based on a subset of participating education systems for which sufficiently complete data were available across student, home, and teacher questionnaires. This restriction, though necessary to ensure data quality and minimize bias due to missing responses, suggests that the findings of this study should not be generalized to the full set of TIMSS participating education systems. It remains possible that the latent class structure or the strength of contextual influences may have differed if the analysis was conducted on a full set of data. Second, while the study identifies associations between parental and teacher engagement and student outcomes, the cross-sectional nature of TIMSS data rules out any causal inferences. Counterfactual analyses, such as evaluating how students with similar characteristics might have engaged under different home or teaching conditions, were not feasible within the scope of this study due to data design constraints. Future research would benefit from longitudinal or quasi-experimental designs to more rigorously examine these dynamics.

In addition, while the selected TIMSS items offered useful entry points for capturing aspects of sustainability engagement, they were not originally designed to comprehensively measure GreenComp competences. Certain GreenComp areas, particularly those emphasizing systems thinking, political agency, or future-oriented reasoning, were underrepresented in this study. As international assessment programs consider ways to integrate sustainability monitoring into their frameworks, intentional alignment with competence-based models like GreenComp would allow for more comprehensive, rigorous, and culturally inclusive exploration.

Despite these constraints, this chapter demonstrates the feasibility and value of repurposing existing international large-scale assessment data available in TIMSS 2023 to inform policy and practice in sustainability education. The association of the GreenComp framework to TIMSS data provides a promising approach for advancing educational monitoring in this emerging domain. By offering an empirically grounded perspective on how young learners engage with sustainability, the study aims to contribute to ongoing policy dialogues, support curriculum development, and underscore the importance of fostering the knowledge, values, and behaviors essential for a sustainable future.

References

Ardoin, N. M., Bowers, A. W., & Gaillard, E. (2019). Environmental education outcomes for conservation: A systematic review. *Biological Conservation, 241*, 108224. https://doi.org/10.1016/j.biocon.2019.108224

Barr, S. (2007). Factors influencing environmental attitudes and behaviors: A U.K. case study of household waste management. *Environment and Behavior, 39*(4), 435–473. https://doi.org/10.1177/0013916505283421

Bianchi, G., Pisiotis, U., & Cabrera Giraldez, C. (2022). *GreenComp: The European sustainability competence framework*. European Commission: Joint Research Centre. Publications Office of the European Union. https://doi.org/10.2760/13286

Boeve-de Pauw, J., Gericke, N., Olsson, D., & Berglund, T. (2015). The effectiveness of education for sustainable development. *Environmental Education Research, 21*(5), 649–679. https://doi.org/10.3390/su71115693

Bookbinder, A., Reynolds, K. A., Leng, D., Tyack, L., Khorramdel, L., & Bezirhan, U. (2025). *Environmental awareness in TIMSS 2023: Patterns in achievement, attitudes, behaviors, and contexts*. TIMSS & PIRLS International Study Center, Boston College. https://doi.org/10.6017/lse.tpisc.qx7218

Brundiers, K., Barth, M., Cebrián, G., Cohen, M., Diaz, L., Doucette-Remington, S., Dripps, W., Habron, G., Harré, N., Jarchow, M., Losch, K., Michel, J., Mochizuki, Y., Rieckmann, M., Parnell, R., Walker, P., & Zint, M. (2021). Key competencies in sustainability in higher education—Toward an agreed-upon reference framework. *Sustainability Science, 16*(1), 13–29. https://doi.org/10.1007/s11625-020-00838-2

Chawla, L. (1998). Significant life experiences revisited: A review of research on sources of pro-environmental sensitivity. *Journal of Environmental Education, 29*(3), 11–21. https://doi.org/10.1080/00958969809599114

Chawla, L., & Cushing, D. F. (2007). Education for strategic environmental behavior. *Environmental Education Research, 13*(4), 437–452. https://doi.org/10.1080/13504620701581539

Ernst, J., & Theimer, S. (2011). Evaluating the effects of environmental education programming on connectedness to nature. *Environmental Education Research, 17*(5), 577–598. https://doi.org/10.1080/13504622.2011.565119

European Commission. (2008). *The Western Balkans in transition* (European Economy, Occasional Papers, No. 46). Directorate-General for Economic and Financial Affairs. https://ec.europa.eu/economy_finance/publications/pages/publication15155_en.pdf.

European Commission. (2019). The European Green Deal. COM(2019) 640 final. https://eur-lex.europa.eu/legal-content/EN/TXT/?uri=celex:52019DC0640.

Evans, G. W., Otto, S., & Kaiser, F. G. (2018). Childhood origins of young adult environmental behavior. *Psychological Science, 29*(5), 679–687. https://doi.org/10.1177/0956797617741894

Evans, N., Whitehouse, H., & Hickey, R. (2012). Pre-service teachers' conceptions of education for sustainability. *Australian Journal of Teacher Education, 37*(7), 1–20. https://doi.org/10.14221/ajte.2012v37n7.3

Franzen, A., & Vogl, D. (2013). Two decades of measuring environmental attitudes: A comparative analysis of 33 countries. *Global Environmental Change, 23*(5), 1001–1008. https://doi.org/10.1016/j.gloenvcha.2013.03.009

Grusec, J. E., & Davidov, M. (2007). Socialization in the family. The roles of parents. In J. E. Grusec & P. D. Hastings (Eds.), *Handbook of socialization: Theory and research* (pp. 284–303). Guilford Press.

Grusec, J. E., & Hastings, P. D. (Eds.). (2015). *Handbook of socialization: Theory and research* (2nd ed.). Guilford Press.

Lee, K., Gjersoe, N., O'Neill, S., & Barnett, J. (2020). Youth perceptions of climate change: A narrative synthesis. *Wiley Interdisciplinary Reviews: Climate Change, 11*(3), e641. https://doi.org/10.1002/wcc.641

Lezhnina, O., & Kismihók, G. (2022). Latent class cluster analysis: selecting the number of clusters. *MethodsX, 9*, 101747. https://doi.org/10.1016/j.mex.2022.101747

Linzer, D. A., & Lewis, J. B. (2011). poLCA: An R package for polytomous variable latent class analysis. *Journal of Statistical Software, 42*(10), 1–29. https://doi.org/10.18637/jss.v042.i10

Little, R. J. A., & Rubin, D. B. (2019). *Statistical analysis with missing data* (3rd ed.). Wiley. https://doi.org/10.1002/9781119482260

Lund, L., & Ritz, C. (2025). Flexible and modular latent transition analysis—A tutorial using R. *PLoS One, 20*(1), e0317617. https://doi.org/10.1371/journal.pone.0317617

Mullis, I. V. S., Martin, M. O., Foy, P., Kelly, D. L., & Fishbein, B. (2020). *TIMSS 2019 international results in mathematics and science*. TIMSS & PIRLS International Study Center, Boston College. https://timssandpirls.bc.edu/timss2019/international-results/

Mullis, I. V. S., Martin, M. O., & von Davier, M. (Eds.). (2021). *TIMSS 2023 assessment frameworks*. TIMSS & PIRLS International Study Center, Boston College. https://timssandpirls.bc.edu/timss2023/frameworks/index.html

Nylund, K. L., Asparouhov, T., & Muthén, B. O. (2007). Deciding on the number of classes in latent class analysis and growth mixture modeling: a Monte Carlo simulation study. *Structural Equation Modeling: A Multidisciplinary Journal, 14*(4), 535–569. https://doi.org/10.1080/10705510701575396

OECD. (2017). *PISA 2015 technical report*. OECD. https://www.oecd.org/content/dam/oecd/en/about/programmes/edu/pisa/publications/technical-report/PISA2015_TechRep_Final.pdf

OECD. (2024). *Western Balkans Competitiveness Outlook 2024: Regional profile, competitiveness and private sector development*. OECD. https://doi.org/10.1787/170b0e53-en

Olsson, D., Gericke, N., & Chang Rundgren, S.-N. (2016). The effect of implementation of education for sustainable development in Swedish compulsory schools. *Environmental Education Research, 22*(2), 176–202. https://doi.org/10.1080/13504622.2015.1005057

Otto, S., & Pensini, P. (2017). Nature-based environmental education of children: Environmental knowledge and connectedness to nature as mediators of pro-environmental behavior. *Global Environmental Change, 47*, 88–94. https://doi.org/10.1016/j.gloenvcha.2017.09.009

Piaget, J. (1972). *The psychology of the child*. Basic Books.

R Core Team. (2019). *R: A language and environment for statistical computing*. R Foundation for Statistical Computing. https://www.R-project.org/.

Ripley, B. D. (2022). *nnet: Feed-forward neural networks and multinomial log-linear models* (R package version 7.3-18). https://CRAN.R-project.org/package=nnet.

Rutkowski, L., & Rutkowski, D. (2010). Getting it 'better': The importance of improving background questionnaire data. *Journal of Curriculum Studies, 42*(3), 411–430. https://doi.org/10.1080/00220272.2010.487546

Schlomer, G. L., Bauman, S., & Card, N. A. (2010). Best practices for missing data management in counseling psychology. *Journal of Counseling Psychology, 57*(1), 1–10. https://doi.org/10.1037/a0018082

Shove, E. (2010). Beyond the ABC: Climate change policy and theories of social change. *Environment and Planning A, 42*(6), 1273–1285. https://doi.org/10.1068/a42282

Stern, P. C. (2000). Toward a coherent theory of environmentally significant behavior. *Journal of Social Issues, 56*(3), 407–424. https://doi.org/10.1111/0022-4537.00175

Stevenson, K. T., Peterson, M. N., Bondell, H. D., Mertig, A. G., & Moore, S. E. (2013). Environmental, institutional, and demographic predictors of environmental literacy among middle school children. *PLoS One, 9*(3), e91031. https://doi.org/10.1371/journal.pone.0059519

Tabachnick, B. G., & Fidell, L. S. (2013). *Using multivariate statistics* (6th ed.). Pearson.

UNESCO. (2017). *Education for sustainable development goals: Learning objectives*. UNESCO. https://doi.org/10.54675/CGBA9153

UNESCO. (2020). *Education for sustainable development: A roadmap*. UNESCO. https://doi.org/10.54675/YFRE1448

UNESCO. (2021). *Reimagining our futures together: A new social contract for education*. UNESCO. https://doi.org/10.54675/ASRB4722

United Nations. (2015). *Transforming our world: The 2030 Agenda for Sustainable Development* (A/RES/70/1). https://sdgs.un.org/2030agenda.

United Nations Economic Commission for Europe. (2012). *Learning for the future: Competences in education for sustainable development*. https://unece.org/info/Environment-Policy/Education-for-Sustainable-Development/pub/3098.

von Davier, M., Fishbein, B., & Kennedy, A. (Eds.). (2024). *TIMSS 2023 technical report (methods and procedures)*. TIMSS & PIRLS International Study Center, Boston College. https://timss2023.org/methods

Vygotsky, L. S. (1978). *Mind in society: The development of higher psychological processes*. Harvard University Press. https://doi.org/10.2307/j.ctvjf9vz4

Wals, A. E. J. (2015). *Beyond unreasonable doubt. Education and learning for socio-ecological sustainability in the Anthropocene*. Wageningen University. https://edepot.wur.nl/365312

Wiek, A., Withycombe, L., & Redman, C. L. (2011). Key competencies in sustainability: A reference framework for academic program development. *Sustainability Science, 6*(2), 203–218. https://doi.org/10.1007/s11625-011-0132-6

Yang Hansen, K., & Gustafsson, J. E. (2016). Determinants of country differences in effects of parental education on children's academic achievement. *Large-Scale Assessments in Education, 4*(11). https://doi.org/10.1186/s40536-016-0027-1

Zelezny, L. C., Chua, P. P., & Aldrich, C. (2000). Elaborating on gender differences in environmentalism. *Journal of Social Issues, 56*(3), 443–457. https://doi.org/10.1111/0022-4537.00177

Paulína Koršňáková is a Senior Research and Liaison Adviser to IEA, and facilitates research networking, stakeholder engagement, and collaboration that enhance capacity building and knowledge sharing to foster innovation and quality in education. She serves as a series editor for the IEA Research for Educators and IEA Teacher Snippets. She holds a master degree in Teacher Preparation Programme of Biology and Mathematics and a doctorate in general psychology.

Tahira Ali Qadri is a Research Officer at IEA in the Netherlands. She specializes in the implementation and quality assurance of international large-scale educational assessments. Her expertise includes quantitative research methods and survey design across education, labour markets and public policy. She holds an MSc in Economics from Universität Hamburg, with a focus on public policy evaluation and applied econometrics.

Wangqiong Ye is a postdoctoral research fellow at the Centre for Research on Equality in Education (CREATE) at the University of Oslo. Her work centers on reducing achievement gaps and promoting educational equity. She employs large-scale assessment data, including TIMSS, PIRLS, and PISA, as well as Norwegian register data. Her methodological expertise spans quantitative approaches such as structural equation modeling (SEM), multilevel modeling (MLM), latent class analysis (LCA), and meta-analysis.

Chapter 4
Attaining Knowledge on Environmental Sustainability: Knowledge and Understanding of Environmental Sustainability Revealed by TIMSS 2023 Data in European Education Systems

Barbara Japelj Pavešić and Jure Novak

4.1 Introduction

Sustainable development has become a critical priority for contemporary societies. The United Nations' 2030 Agenda established goals towards sustainable development that address social, economic, and environmental challenges, including ending hunger and poverty, protecting natural resources, promoting sustainable production and consumption, and reducing inequality (United Nations, 2015). One highly effective approach to achieving these goals is sustainable education (Krayneva et al., 2021), as emphasized by the United Nations Educational, Scientific and Cultural Organization (UNESCO) in *Education for sustainable development: A roadmap* (UNESCO, 2020). Within this framework, environmental sustainability is a key domain and is the central focus of this chapter and volume.

The topic of environmental sustainability is most commonly integrated into the science curriculum in schools. Therefore, examining learning opportunities and outcomes in science education in schools across different regions can provide critical perspectives into environmental sustainability education.

The research literature offers valuable insights into education for environmental sustainability and the teaching of specific environmental topics across different education systems and educational levels. Education systems often compare their approaches and practices when introducing new strategies, as seen in a comparative study of education for sustainable development in Croatia, Montenegro and Serbia (Vukić et al., 2021). However, there is limited evidence on students' environmental knowledge, particularly in the early years of education. For instance, a study from Momete and Momete (2021) attempting to develop a method for tracking progress

B. J. Pavešić (✉) · J. Novak
Educational Research Institute, Ljubljana, Slovenia
e-mail: barbara.japelj@pei.si; Jure.novak@pei.si

D. Hastedt et al. (eds.), *Educating Pupils for Environmental Sustainability in Europe*, IEA Research for Education 18,
https://doi.org/10.1007/978-3-032-14725-7_4

in education for sustainable development across the European Union, focusing on how education systems are meeting Sustainable Development Goals, included information about environmental knowledge. European Union education systems were grouped into higher and lower achievers based on educational indicators. However, the primary knowledge indicators used did not specifically assess sustainability knowledge, competencies, or attitudes. Instead, they measured the general science competencies of 15-year-old students as reported in the Organisation for Economic Co-operation and Development's PISA[1] study (Momete & Momete, 2021). A recent systematic review of studies on environmental sustainability competencies in primary education (Cebrián et al., 2025) examined existing approaches and tools assessing sustainability competencies but did not address student knowledge directly. The review identifies a range of pedagogical strategies used to assess knowledge, competencies and outcomes, primarily within interdisciplinary environmental learning activities for students in different age groups. It highlights the need for greater evidence to inform educators about the current state of education for environmental sustainability with regard to student knowledge and competences (Cebrián et al., 2025).

Key predictors of environmentally responsible behavior include attitudes and social norms, both of which are expected to be closely linked to environmental knowledge, according to educational research (Reynolds & Komakhidze, 2022). Increased knowledge can enhance motivation for engagement, while active participation can, in turn, deepen knowledge. The overarching goal of education for environmental sustainability is to equip students—and future adults—with the knowledge and motivation needed to take action in preserving the environment and promoting sustainability. Consequently, the relationship between knowledge acquired in school and the development of attitudes is crucial. Baierl et al. (2024) found that students' attitudes toward nature influenced the levels of knowledge they gained and retained. Earlier research also demonstrated that students' motivation for environmental engagement varies according to age and school grade level (Liefländer & Bogner, 2014). To develop effective strategies for fostering long-term motivation, it is essential to understand how knowledge shapes engagement and how these dynamics differ across education systems and cultural contexts from an early age.

In Chaps. 1 and 2 of this volume, existing and emerging opportunities to learn environmental sustainability are examined through GreenComp, the reference framework for sustainability competences in the European Union, alongside additional frameworks available across Europe (Bianchi et al., 2022). Understanding the relationship between knowledge and attitudes—while considering the opportunities students have to learn—represents a crucial next step in developing effective strategies to enhance sustainability education and promote a deeper understanding of environmental issues.

The central purpose of this chapter is to explore how education for environmental sustainability is implemented in European education systems, based on research of student knowledge. In 2023, most European education systems took part in the Trends

[1] PISA = Programme for International Student Assessment.

in International Mathematics and Science Study (TIMSS) 2023, which assessed students' environmental knowledge and attitudes, and factors influencing their learning opportunities. TIMSS is a complex curriculum-based study which collects evidence from multiple sources while maintaining connections between them, enabling the interpretation of all collected data to the measured student achievement within representative samples from each participating education system.

4.1.1 The TIMSS Curriculum Model Behind the Data on Education for Environmental Sustainability

Among international large-scale assessments, TIMSS 2023 is the most extensive curriculum-based study of science knowledge (von Davier, Kennedy, et al., 2024). Conducted in 4-year cycles, TIMSS measures students' knowledge, based on a selected set of contents and topics that students in all participating education systems are generally expected to have the opportunity to learn in school (achievement data). In addition to achievement data, TIMSS collects contextual data to explain differences in students' learning opportunities and measured knowledge. The contextual data is gathered from multiple sources—including national representatives, schools, teachers and students—through questionnaires and national reports. The focus of this chapter is specifically on environmental achievement and curricular data for the target population of grade four students in European education systems.

4.1.2 Contents of Environmental Achievement Tests

For each study cycle, the participating education systems together select contents and topics that form the basis of the common achievement tests. These are documented in the TIMSS assessment frameworks (Mullis et al., 2021) and are used to develop the assessment items that make up the student achievement tests. In TIMSS 2023, there were 44 assessment items aligned with 14 environmental sustainability topics (Table 4.1).

4.1.3 Curricular Data

TIMSS is structured around three curricular levels: the intended curriculum, the implemented curriculum, and the attained curriculum (Fig. 4.1). At each level, different data on environmental sustainability are collected in a way that allow them to be related to student achievement, which is the focus of the study.

Table 4.1 Environmental contents assessed in TIMSS 2023

Content	Topics
Earth science	
ES/ ECH/1A	Physical characteristics of the Earth system: Recognize that Earth's surface is made up of land and water in unequal proportions (more water than land) and is surrounded by air; describe where fresh and salt water are found
ES/ ECH/2A	Earth's resources: Identify some of Earth's resources that are used in everyday life (e.g., water, wind, soil, forests, oil, natural gas, minerals)
ES/ ECH/2B	Earth's resources: Explain the importance of using Earth's renewable and nonrenewable resources responsibly (e.g., fossil fuels, forests, water)
ES/ ECH/3B	Earth's history: Recognize that some remains (fossils) of animals and plants that lived on Earth a long time ago are found in rocks and ice and make simple deductions about changes in Earth's surface from the location of these remains
ES/ ECH/1B	Weather and climates on Earth: Describe how weather (i.e., daily variations in temperature, humidity, precipitation in the form of rain or snow, clouds, and wind) can vary with geographic location
ES/ ECH/1C	Weather and climates on Earth: Describe how average temperature and precipitation can change with the seasons and location; recognize that the average temperature on Earth has increased over the last century and some effects of this increase on Earth's physical characteristics (e.g., ocean levels have increased, ice caps have melted, rivers have dried up, deserts have grown bigger)
Life science	
LS/ Eco/1A	Common ecosystems: Relate common plants and animals (e.g., evergreen trees, frogs, lions) to common ecosystems (e.g., forests, ponds, grasslands)
LS/ Eco/2A	Relationships in simple food chains: Recognize that plants need (sun)light, air, and water to provide energy for life processes (i.e., growth and repair, movement, and reproduction); explain that animals eat plants or other animals to get the food they need to supply energy for life processes (i.e., growth and repair, movement, and reproduction)
LS/ Eco/2B	Relationships in simple food chains: Complete a model of a simple food chain using common plants and animals from common ecosystems, (e.g., a forest, a desert, a river, an ocean)
LS/ Eco/2C	Relationships in simple food chains: Describe the roles of living things at each link in a simple food chain (e.g., plants produce their own food; some animals eat plants, while other animals eat the animals that eat plants)
LS/ Eco/2D	Relationships in simple food chains: Identify common predators and their prey and describe their relationships
LS/ Eco/3A	Competition in ecosystems: Recognize and explain that some living things in an ecosystem compete with others for resources (e.g., food, light, space)
LS/ OEI/3A	The impact of humans on the environment: Recognize that human behavior has negative and positive effects on the environment (e.g., negative effects of air and water pollution, positive effects of reducing air and water pollution); provide general descriptions and examples of the effects of pollution on humans, plants, and animals
Physical science	
PS/ EN/1A	Common sources and uses of energy: Identify sources of energy (e.g., the Sun, flowing water, wind, coal, oil, gas), and recognize that energy is needed for movement and transportation, manufacturing, heating, lighting, and powering electronic devices

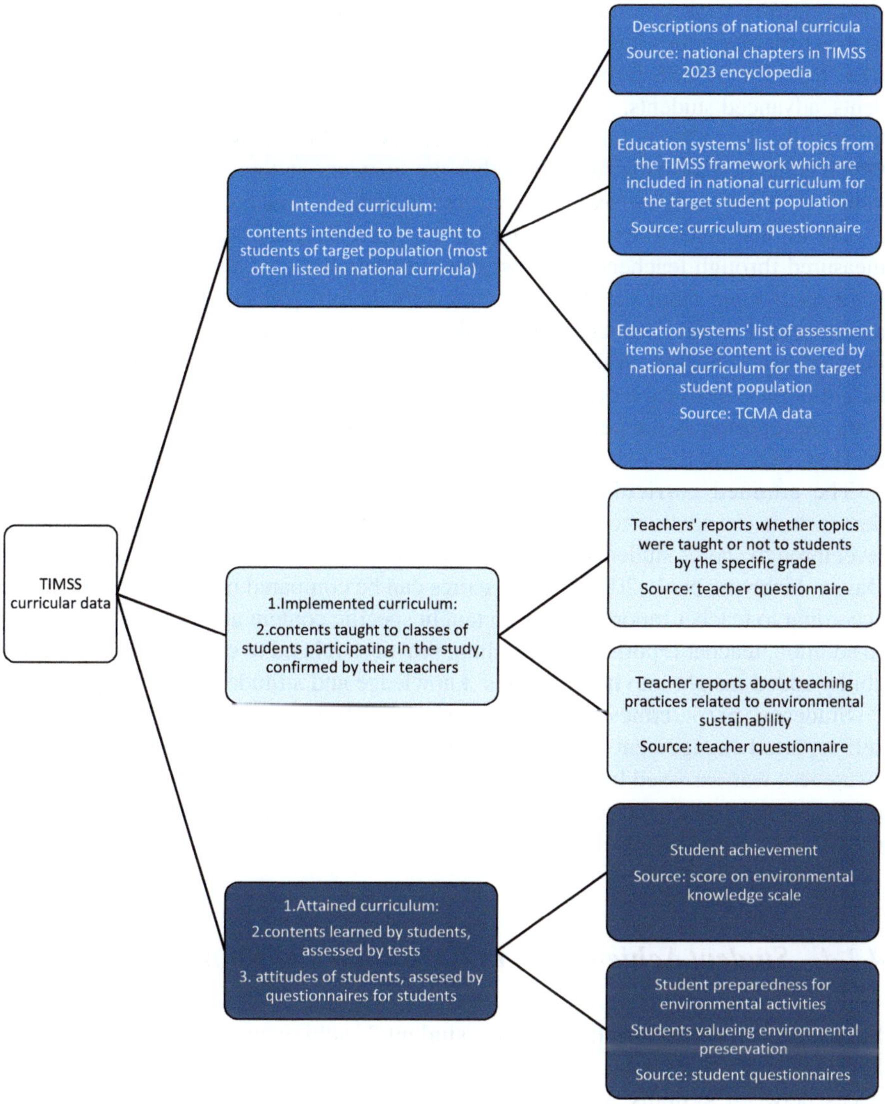

Fig. 4.1 Model of data sources for teaching the environmental sustainability contents

The intended curriculum refers to the topics intended to be taught in schools and is based on information collected on the national level. Each participating education system is asked to provide a national chapter to the TIMSS 2023 encyclopedia (Reynolds et al., 2024), presenting an overview of their mathematics and science education. The chapters follow a similar structure and, among other information, provide summaries of the science curriculum up to grade four and grade eight. These summaries offer a broad perspective on the national curricula, beyond the specific environmental sustainability topics examined in TIMSS. Further, education systems' responses to the curriculum questionnaire (Reynolds et al., 2024), report whether each specific topic from the TIMSS framework is included in the national curriculum for all students, advanced students, or not included at all. More precise data are also collected through the Test Curriculum Matching Analysis (TCMA) (Fishbein et al., 2025), in which education systems report whether the contents or topics of each assessment item used in the study are explicitly covered in their national curricula.

The implemented curriculum refers to the topics actually taught in schools and is measured through teachers' responses on whether students had the opportunity to learn each topic from the TIMSS framework before participating in the TIMSS assessment. For environmental sustainability topics, science teachers of participating students indicated, in their teacher questionnaire, whether each topic had been taught in previous years, during the year of assessment, or had not yet been taught. Data are reported as the percentage of students who have been taught each topic in previous years, in the assessment year, or not at all, by education systems.

The attained curriculum refers to what students actually learned and can demonstrate. In relation to environmental sustainability, it is observed at the student level through scores students achieved on the scale of environmental knowledge (von Davier, Fishbein, et al., 2024). These scores can be compared between students who, according to teacher reports, had been taught specific content and those who had not. In addition, teacher reports about teaching practices related to environmental sustainability added insight into how students' knowledge and attitudes are developed.

Students' preparedness for environmental activities was evaluated based on self-reported engagement in environmentally responsible behaviors. Environmental awareness was assessed by asking students about their feelings toward nature, with responses used to construct the Students Value Environmental Preservation scale (von Davier, Kennedy, et al., 2024).

4.1.4 Student Achievement in Environmental Knowledge

Focusing on student achievement, students' achievement in environmental knowledge can be examined by education system. This reveals relatively large differences between systems, with no clear geographic pattern apparent on the scale (Table 4.2).

Table 4.2 Mean student score on the scale of environmental knowledge

Education system	Mean	SE	Education system	Mean	SE
Poland	556.91	(2.56)	Serbia	517.03	(3.26)
Finland	543.10	(3.37)	Germany	512.85	(3.19)
Ireland	536.17	(3.94)	Spain	510.15	(2.64)
Denmark	535.70	(3.58)	Slovak Republic	509.00	(4.46)
Hungary	533.07	(4.24)	Latvia	503.78	(3.48)
Bulgaria	532.60	(5.84)	France	495.70	(4.03)
Sweden	527.70	(3.81)	Belgium (Flemish)	491.70	(3.56)
Romania	525.40	(5.97)	Belgium (French)	482.74	(3.09)
Lithuania	520.47	(3.32)	Cyprus	479.67	(3.18)
Czech Republic	520.15	(3.07)	Albania	478.18	(6.25)
Slovenia	520.02	(3.03)	Montenegro	465.20	(2.59)
Netherlands	518.91	(3.05)	Bosnia and Herzegovina	448.45	(5.07)
Portugal	518.39	(3.64)	N. Macedonia, Rep. of	436.86	(5.64)
Italy	517.06	(2.72)	Kosovo	402.39	(4.08)
Table average: 504.98 (0.75)					

Note: SE, standard error

4.1.5 *Aim of the Study*

The aim of this chapter is threefold. First, it introduces the reader to the complex landscape of TIMSS curricular and achievement data. Second, it presents the results of analyses that explore patterns, relationships and comparisons in environmental sustainability education across European education systems. Finally, the chapter seeks to provide insight into students' opportunities to learn about environmental sustainability, their knowledge and attitudes toward it, and the relationship between these factors.

4.1.6 *Research Questions*

The main research questions addressed in this study are:

1. How do national curricula, teaching approaches, and student outcomes in education for environmental sustainability vary across European education systems?
2. How do the prescribed content and the content taught relate to student outcomes?
3. Is higher student environmental knowledge linked to more positive attitudes toward environmental sustainability?
4. Are there groups of European education systems that share similarities in teaching approaches, student knowledge, and attitudes toward environmental sustainability?

4.2 Data Sources and Methods

The data sources for this study consisted of national reports and three TIMSS 2023 datasets from participating European education systems. Analysis was conducted of TIMSS 2023 encyclopedia chapters, curricular data, and achievement and attitudinal data of grade four students, along with teacher-reported data, across 28 education systems: Albania, Belgium (Flemish), Belgium (French), Bosnia and Herzegovina, Bulgaria, Cyprus, the Czech Republic, Denmark, Finland, France, Germany, Hungary, Ireland, Italy, Kosovo, Latvia, Lithuania, Montenegro, the Netherlands, the Republic of North Macedonia, Poland, Portugal, Romania, Serbia, the Slovak Republic, Slovenia, Spain, and Sweden.

4.2.1 *Intended Curricula*

The analysis of intended curricula involved four steps:

1. Extracting information on environmental sustainability contents in science curricula from the national chapters of the TIMSS 2023 encyclopedia.
2. Reviewing data from the curricular questionnaire to determine which observed environmental sustainability topics were covered for all students, some students, or not at all in national curricula.
3. Analyzing TCMA data to compare education systems by assessment items, with regard to whether their content was covered in national curricula. To identify patterns across education systems, hierarchical clustering using the Ward method was applied, highlighting similarities between systems in including environmental sustainability contents in their curricula.
4. Relating coverage of environmental sustainability content in intended curricula with student mean achievement in environmental knowledge, by education system.

4.2.2 *Implemented Curricula*

The analysis of implemented curricula involved three steps:

1. Comparing the extent of topics taught, as reported by teachers, by education system.
2. Relating information about topics taught with student achievement in environmental knowledge. Student groups were compared based on whether they had been taught the topics before or during the assessment year, or not yet, analyzing differences in group size and mean achievement.

3. Reviewing teacher reports on the use of teaching practices related to environmental sustainability and linking them to student achievement. Patterns of promoting environmental activities by education system were examined:
   (a) Frequency of engagement in specific activities in lessons, such as developing students' positive attitudes toward the natural environment, encouraging resource conservation (for example, water, energy), discussing how students' actions inside and outside of school can help the environment, and addressing environmental issues like climate change and endangered species.
   (b) Involvement of students in activities like visiting natural areas, participating in environmentally responsible actions (for example, picking up trash), conducting research or projects on environmental topics (for example, pollution, climate change), and providing opportunities for students to engage in outdoor environmental education programs outside of school.

4.2.3 Attained Curricula

The analysis of attained curricula focused on student attitudinal and achievement data. TIMSS was designed to assess students' attitudes toward the natural environment and their engagement in environmentally responsible behaviors. Students were asked two sets of questions: how much they agree with specific statements about their relationship with the natural environment, and how frequently they take actions to help it. The responses to the first set of statements were used to create the Students Value Environmental Preservation scale, with an international mean score of 10 points. The second set of questions provided data on the frequency of students' environmental activities.

To explore the connection between attitudes toward environmental sustainability and environmental knowledge, the following analyses were conducted:

- Correlation analyses between environmental knowledge and student attitudes toward environment.
- A regression analysis of the Students Value Environmental Preservation scale and achievement in environmental knowledge.
- Comparisons between education systems of students' actual engagement in environmentally friendly activities.

To ensure accurate calculations, TIMSS-prescribed procedures were followed for computing mean student achievement, using the International Association for the Evaluation of Educational Achievement (IEA) IDB (International Database) Analyzer (IEA, 2021) with SPSS (IBM, 2021) to correctly handle plausible values for achievement scales and account for sampling errors (Fishbein et al., 2025).

4.3 Results

4.3.1 Level of Intended Curricula

4.3.1.1 Extracting Information on Environmental Sustainability Content in Science Curricula

The review of national science curricula from the TIMSS 2023 encyclopedia (Reynolds et al., 2024) provides a general overview of intended curricula for teaching environmental sustainability (Table 4.3). There is a notable degree of variation in education systems' reports. In some systems, topics are allocated to different grades, others report only the general contents requested by grade four.

Table 4.3 Summary of content prescribed in national curricula related to environmental awareness and sustainability

Education system[a]	Summary from TIMSS 2023 encyclopedia on environmental awareness and sustainability topics included in national curricula
Poland	The core curriculum emphasizes the need to organize activities that support perceiving the natural environment and exploring it, learning the values and inter-relationships of components of the natural environment, learning the values and norms that source a healthy ecosystem, and developing behaviors resulting from these values
Finland	*Building a Sustainable Future* unit: Students study biodiversity, climate change, and its mitigation; sustainable use of natural resources; promoting health; caring for one's own cultural heritage; living in a multicultural world; and the global well-being of humanity now and in the future. Students take part in a collaborative project in which they practice participation and involvement at the local or global level
Ireland	*Environmental Awareness and Care* is one of four content strands, subdivided into three units *Environmental Awareness*: identify positive aspects of natural and built environments; identify the interrelationship of living and nonliving elements of local and other environments; become aware of the importance of Earth's renewable and nonrenewable resources; recognize how people's actions affect their environment; come to appreciate the need to conserve resources *Science and the Environment*: begin to explore and appreciate the application of science and technology in familiar contexts; identify some ways in which science and technology contribute positively to society; recognize and investigate human activities that have positive and negative effects on local and wider environments *Caring for the Environment*: examine a number of ways in which the local environment could be improved or enhanced; identify and discuss a local, national, or global environmental issue; realize that there is a personal and community responsibility for taking care of the environment
Denmark	The science subject should develop students' understanding of human interaction with nature and encourage them to engage and act in ways that support sustainable development
Hungary	*Integrated Natural Sciences* subject: explaining the importance of a sustainable lifestyle by using examples, interpreting the role of tradition in building a harmonic connection with nature

(continued)

Table 4.3 (continued)

Education system[a]	Summary from TIMSS 2023 encyclopedia on environmental awareness and sustainability topics included in national curricula
Bulgaria	Identify common environmental pollutants; identify human actions that can affect the balance of nature and describe possible courses of action for environmental protection
Sweden	Grades four to six: *Nature and Environment* as part of core content—human dependence on and impact on nature, with links to the use of natural resources, sustainable development, and ecosystem services; nature as a resource and our responsibilities when using it
Romania	Grades three to four: General aim—develop an interest in and feel responsible for environmental sustainability; curriculum knowledge standard—Basic competencies in science and technology: exhibit interest in one's own health and in a clean environment; apply basic rules for personal hygiene and rules for responsible behavior toward the environment
Lithuania	*Humans and Nature* content area: recognize that the environment and natural phenomena (sunlight, air, and water) help sustain life
Czech Republic	*Diversity of Nature* thematic area with *Conservation and protection* topic: human responsibility toward the environment, conservation, and protection of the environment; waste disposal; natural and ecological disasters
Slovenia	Grades one to three: environmental pollution, consequences of pollution on living things, waste management, pollutants of water, soil, air, energy saving, environmental management
Netherlands	The eight core objectives are focused on natural science, technology, and health-related topics. They provide a general overview of skills and knowledge, including having an investigative attitude toward nature, connecting theories and models with practical work, and promoting sustainability
Portugal	One of six themes: discovering relations between nature and society
Italy	Grade three: *Man, Living Things, and the Environment* content area—recognize needs in different living organisms, similar to those of humankind, with respect to their environment; and, *Observing and Experimenting in the Field* content area—observe and understand the environmental changes generated by nature (e.g., sunlight, weathering, water) and by humankind (e.g., urbanization, cultivation, industrialization)
Serbia	Grade three: Expected outcomes—illustrate with examples responsible and irresponsible attitudes of humans toward the environment; explain how recycling helps to preserve nature
Germany	*Environment and Sustainability* topic: identify possibilities for waste prevention and prepare a guidebook for this purpose; investigate and discuss the importance and use of resources and test their careful use (e.g., water, energy, soil, air, paper)
Spain	*The Living Things* content block with topics: relationship between living things; characteristics and components of an ecosystem From the 2023–2024 academic year, *Knowledge of the Natural, Social, and Cultural Environment* subject with topic: identify the causes and consequences of human intervention in the environment from social, economic, cultural, technological, and environmental perspectives in order to improve one's ability to face problems and to implement sustainable lifestyles that are consistent with respect for, care for, and the protection of people and the planet

(continued)

Table 4.3 (continued)

Education system[a]	Summary from TIMSS 2023 encyclopedia on environmental awareness and sustainability topics included in national curricula
Slovak Republic	Under theme *Plants*, the topics: germination, vegetative reproduction, conditions for plant growth, effects of water and light on plants, diversity of plants, pollinators, lifespan of plants, effects of environmental changes
Latvia	*Interaction of Man and Environment* subject, with topics: Learn to identify protected natural objects in the nearest vicinity, acknowledge experience to care for living beings and be aware of the necessity of taking care of living beings, observe the norms of behavior in nature (e.g., protected areas), learn to safely collect mushrooms and herbs New (2022): What are natural resources and how are they used? Raise awareness of renewable and non-renewable natural resources, how they can be used, and the need to conserve them
France	Grades four to six: Each of four main themes, (1) *Matter, Movement, Energy, Information*; (2) *Living Things, Their Diversity, and Their Functions*; (3) *Technical Materials and Objects*; and (4) *Planet Earth—Living Beings in Their Environment* enables the construction of concepts or notions that find their application in education for sustainable development. The theme *Planet Earth—Living Beings in Their Environment* includes the topic: identify issues related to the environment
Belgium (Flemish)	*Nature* domain: The students can independently perform basic actions in the care of animals and plants in their environment; The students show in their behavior a willingness to handle waste, energy, paper, food, and water carefully in their class and school; The students can illustrate with concrete examples from their environment how people deal with the environment positively and negatively; The students can illustrate with concrete examples from their environment that conflicting interests often underlie environmental problems; The students show respect and care for nature, understanding that human needs depend on the natural environment
Belgium (French)	Knowledge domain of *People and the Environment*: The main objective is to act in an informed manner for the benefit of all people and science teaching contributes to the development of skills. Environmental education (management, conservation, protection, use of resources and destruction, pollution, etc.) should not be certified but rather focused on constantly raising awareness
Cyprus	*Life Science* subject: Cyprus ecosystems and the importance of their conservation
Albania	Sustainability contents are not explicitly mentioned
Montenegro	*Life cycles* topics: understand the correlation between people's activities and pollution; understand man's impact on water conservation and consumption; understand the importance of protecting the air
Bosnia & Herzegovina	Sustainability contents are not explicitly mentioned
N. Macedonia, Rep. of	Natural resources and their protection (natural resources, air, water, soils, forests, mineral resources, coal, oil, natural/terrestrial gas, sun, water, wind, non-renewable sources of energy, renewable energy sources, recycling)
Kosovo	Sustainability contents are not explicitly mentioned

Note: [a]Education systems are ordered according to the national mean student achievement in environmental knowledge; see Table 4.2. If a grade is not mentioned, it is reporting grade four results

4.3.1.2 Reviewing Data from the Curricular Questionnaire: Topics from the TIMSS Framework Intended to be Taught

The analysis of coverage of observed environmental sustainability contents from the TIMSS framework in national curricula shows more comparable results (Tables 4.4 and 4.5). For some topics, education systems reported combined coverage of sets of topics.

Summaries of coverage of environmental sustainability content in national curricula reveal distinct approaches across education systems to integrating environmental topics into curricula. While a few systems do not explicitly mention environmental content in the TIMSS 2023 encyclopedia, nearly all education systems emphasize teaching the importance of care for the natural environment, understanding human impact on nature, addressing pollution and waste management, and fostering positive attitudes toward sustainability. Although the term *sustainability* is not explicitly used in many national reports, it is likely incorporated in school teaching, given that the reports are summaries and may not capture all details.

Regarding content and topics from the TIMSS framework, some are intended to be taught in most education systems, at least to advanced students (for example, positive and negative impacts of humans on the environment), while others appear in only a few systems (for example, competition in ecosystems, and fossils and

Table 4.4 Life and physical science topics reported by education systems as included in curriculum

Topics as reported by education systems[a]	The positive and negative impacts of humans on the environment; LS/OEI/3A	Plants and animals in common ecosystems; LS/Eco/1A	Relationships in simple food chains; LS/Eco/2A, LS/Eco/2B, LS/Eco/2C, LS/Eco/2D	Competition in ecosystems; LS/Eco/3A	Common sources of energy; PS/EN/1A
Poland	Advanced	None	All	None	None
Hungary	All	All	None	None	All
Finland	All	None	All	None	None
Slovenia	All	All	None	None	None
Belgium (Flemish)	Advanced	Advanced	Advanced	None	Advanced
– Education systems where all observed topics are intended to be taught to all students: Albania, Bulgaria, Denmark, France, Germany, Italy, Kosovo, Latvia, Lithuania, Montenegro, Portugal, Romania, Serbia, Spain, and Sweden					
– Education systems where competition in ecosystems is not intended to be taught to any students and four other topics to all students: Belgium (French), Bosnia and Herzegovina, Cyprus, the Czech Republic, Ireland, the Republic of North Macedonia, and the Slovak Republic					

Notes. [a]Education systems are ordered according to the national mean student achievement in environmental knowledge; see Table 4.2. All: taught to all students; advanced: taught to advanced students only; none: not taught. The Netherlands was excluded from the analysis

Table 4.5 Earth science topics reported by education systems as included in curriculum

Topics as reported by education systems[a]	Physical characteristics of Earth; ES/ ECH/1A	Earth's renewable and non-renewable resources; ES/ ECH/2A, ES/ ECH/2B	How weather can vary across geographic locations and seasons; ES/ ECH/1B	Earth's rising average temperatures and results of this change; ES/ ECH/1C
Finland	None	None	All	None
Ireland	None	All	None	None
Hungary	All	All	None	None
Czech Republic	All	All	All	Advanced
Slovenia	All	None	All	None
Germany	All	All	All	Advanced
Spain	All	None	All	None
Slovak Republic	All	None	All	None
France	All	All	None	All
Belgium (Flemish)	All	Advanced	All	Advanced
Belgium (French)	All	None	All	None
Bosnia and Herzegovina	All	None	All	Advanced
N. Macedonia, Rep. of	All	All	None	None
Kosovo	All	All	All	None

– Education systems where all observed topics, including fossils, are intended to be taught to all students: Denmark, Italy, and Romania
– Fossils and what they show about Earth's history are not intended to be taught to any students in all other education systems
– All four topics except fossils and what they show about Earth's history were taught in Latvia, Lithuania, Portugal, Serbia and Sweden
– All three topics except fossils and Earth's rising average temperatures and results of this change are taught in Bulgaria, Cyprus, Kosovo and Montenegro
– All five topics were not intended to be taught to any students in Albania and Poland

Notes. [a]Education systems are ordered according to the national mean student achievement in environmental knowledge; see Table 4.2. All: taught to all students; advanced: taught to advanced students only; none: not taught. The Netherlands was excluded from the analysis

what they show about Earth's history). Overall, the Earth science topics are less frequently included in national curricula compared to life and physical science topics. In almost one third of education systems, more than half of the Earth science topics (three or more out of five) are not covered in the national curricula at all.

The least taught topic (after fossils) is Earth's rising average temperature and results of this change, which is directly linked to the environmental sustainability problem of climate change. This topic is taught in only 9 education systems to all students, in 4 education systems to advanced students, but in 14 education systems to no students at all. In Finland and Ireland, general climate change content is explicitly included into their summaries of curriculum, but as a specific topic it is marked as not intended to be taught to the same population of students, revealing the potential differences in transforming national curricula into teaching of specific content.

4.3.1.3 Analyzing Test Curriculum Matching Analysis (TCMA) Data: Assessment Items' Content Coverage in National Curricula

The third and most precise source of information about intended curricula is the national list of assessment items identified as being covered in each national curriculum, meaning that students are intended to be taught the content and concepts that are needed to solve items correctly. Of the 28 education systems, 2 systems—Bulgaria and Cyprus—did not provide data for this analysis. For the remaining 26 education systems, content coverage for each of the 44 items forming the environmental knowledge scale was analyzed. Education systems were then grouped into four distinct clusters based on similarities in intention to teach the contents of the environmental knowledge scale items:

- The largest cluster—with the greatest number of shared scale items intended to be taught—includes half of the education systems: Belgium (French), Bosnia and Herzegovina, Denmark, France, Germany, Ireland, Italy, Lithuania, Montenegro, the Republic of North Macedonia, Portugal, Romania, Serbia, and Sweden.
- The second cluster includes four education systems: the Czech Republic, Latvia, the Netherlands, and Spain.
- The third cluster includes three education systems: Albania, Finland, and the Slovak Republic.
- The fourth cluster—with the fewest number of shared scale items intended to be taught—includes five education systems: Belgium (Flemish), Hungary, Kosovo, Poland and Slovenia.

The assignment of education systems to clusters shows no geographical pattern (for example, Latvia and the Netherlands appear in the same cluster) and no direct relationship to mean student achievement (for example, Albania and Finland appear in the same cluster).

In general, the results align with education systems' reports on coverage of content and topics in national curricula. Systems in the fourth cluster tend to report less intended teaching of environmental topics, as reflected in their curriculum analysis reports (Tables 4.4 and 4.5). For example, Poland reported almost all TIMSS topics were not intended to be taught, while Belgium (Flemish) indicated that almost all topics were intended for advanced students only.

However, not all education systems in this group show consistent results from the two analyses. For example, the Czech Republic reported that almost all TIMSS topics were covered by the curriculum, including sources of energy, yet items asking about the use of these sources of energy were not identified as being intended to be taught. This discrepancy highlights the potential problem of overgeneralization: curriculum documents may specify broad content areas, while the specific concepts tested by individual items are not always intended to be taught. This suggests that while parts of the observed topics might be presented to students, the full scope of the content may not be consistently included, underscoring limitations in the precision of curriculum descriptions and assessments.

Reviewing the item coverage for one education system from each cluster, focusing on items that differ in coverage between clusters, provides further insight into the characteristics of each cluster (Table 4.6).

4.3.1.4 Relations to Student Achievement

The analysis of student knowledge indicated that mean national student achievement in environmental knowledge is not directly related to the number of items covered in the curricula of education systems (Fig. 4.2). In general, the mean achievement of education systems in Cluster 4, which cover the fewest items, is not lower than the mean achievement of education systems in Clusters 2 or 3, which cover a larger number of items and concepts in their curricula. For example, Poland has the highest achievement but covers almost the fewest items. Finland shows coverage of 31 items and achieves a high score of over 500. Italy and the Republic of North Macedonia have the highest coverage of items, yet there is a large difference in student environmental knowledge between the two systems (Fig. 4.2).

The results clearly indicate that simply covering content in the curriculum does not always lead to higher student knowledge.

4.3.2 Level of Implemented Curricula

4.3.2.1 Comparing the Extent of Topics Taught in Schools

Implemented curricula—the contents actually taught in classes—were measured through teachers' reports indicating when topics from the TIMSS framework were taught to their assessed students.

The results reveal that, in many education systems, teachers report different levels of curriculum coverage than national reports. For topics that national reports indicated were not intended to be taught, teachers confirmed teaching them to notable percentages of students (Table 4.7). The opposite pattern was also observed: content listed in national curricula was reported by teachers as not being taught to large fractions of students.

For example, Finland, Poland and Spain reported that their national curricula cover the teaching of relationships in simple food chains while Hungary and Slovenia do not. In practice, teachers in Poland reported that 86% of students were taught this content in the years before grade four and almost all remaining students learned this content in grade four. In contrast, in Spain, almost 40% of

Table 4.6 Items and their content coverage by curricula

Item	Topic	Coverage	LVA	CZE	FIN	POL
SE81168	Why plastic objects dangerous	26	1	1		1
SE81077	Plant in the cupboard dies	26	1	1		1
SE81036	Decrease air pollution in a city	24	1	1	1	
SE81044	Harmful and helpful human activities	24	1	1	1	
SE81202	Location of salt water	24	1	1	1	
SE71071	Mice increase after trees are cut down	24	1	1		1
SE81055	Human activities that help or harm the environment	24	1	1		
SE81920B	Disadvantages to farming near a river	24		1		1
SE81920A	Advantages to farming near a river	24		1		1
SE71201	What covers Earth's surface	23	1	1		
SE61069	Plants make their own food	22	1	1		1
SE81079	Grassland food chain	22	1	1		
SE71063	Animals that live in the desert	20	1	1	1	
SE81080	Animals found in a desert	20	1	1	1	
SE81238	Effects of increase in Earth's temperature	20	1	1		
SE81054	Connection between habitat and ape population	20	1			1
SE81204	Source of water for a desalination plant	20		1	1	
SE81070	Predators and prey on the savanna	19	1	1	1	
SE61116	Temperature graphs	19	1	1	1	
SE81239	Place that with hot, dry weather	19	1	1	1	
SE71081	Animal that competes with giraffe for food	19	1	1	1	
SE61124	Recycle products made of metal	19		1	1	
SE81212	Renewable and non-renewable sources of energy	19			1	
SE81207B	Energy transformed by turbines	18	1		1	
SE81207A	Energy captured by turbines	18	1			
SE71220	Tropical plant fossil in cold environment	10			1	
SE71222	Fish fossil in the desert	9			1	

Notes: CZE, Czech Republic; FIN, Finland; LVA, Latvia; POL, Poland; 1, the item is covered by the education system. Coverage refers to the number of education systems which reported the item's content as covered by national curricula. The coverage of all items by each education system is provided in Table 4.11

students were reportedly not taught this content in schools, according to their teachers. Similarly, teachers in Finland reported that about a quarter of students were not taught the content by the time of the TIMSS assessment. In Hungary, where this content is not covered in the national curriculum, the content was nevertheless taught to 62% of students. In Slovenia, where this content is also not

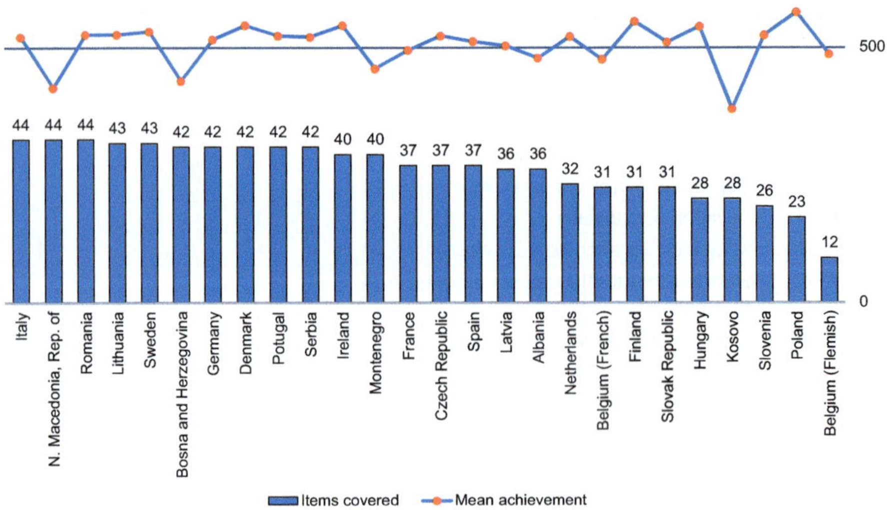

Fig. 4.2 Similar education systems by coverage of items' contents and student environmental knowledge

Table 4.7 Percentage of students being taught two topics by their teachers in selected education systems

Education system	Topics taught in schools	Relationships in simple food chains %	Common sources of energy %	Education system	Topics taught in schools	Relationships in simple food chains %	Common sources of energy %
Belgium (Flemish)	Before	49.3 (5.5)	12.0 (3.2)	Finland	Before	47.7 (3.7)	29.0 (3.3)
	This year	29.2 (4.7)	55.5 (4.7)		This year	24.9 (3.2)	45.6 (3.7)
	Not yet	21.5 (4.0)	32.4 (4.3)		Not yet	27.4 (3.5)	25.3 (3.0)
Hungary	Before	32.0 (3.3)	13.9 (2.3)	Poland	Before	86.1 (2.4)	43.0 (4.0)
	This year	39.5 (3.0)	39.6 (3.6)		This year	13.5 (2.4)	55.4 (3.9)
	Not yet	28.4 (3.2)	46.5 (3.7)		Not yet	0.4 (0.4)	1.6 (0.7)
Slovenia	Before	25.4 (3.8)	21.4 (3.0)	Spain	Before	21.1 (3.3)	11.2 (2.2)
	This year	55.9 (4.5)	72.5 (3.2)		This year	41.0 (3.8)	49.5 (4.4)
	Not yet	18.7 (3.1)	6.2 (1.7)		Not yet	37.9 (3.2)	39.4 (3.8)

Notes: Standard errors appear in parentheses. Due to rounding some values may seem inconsistent. The results for all education systems are in Table 4.16

covered in the national curriculum, teachers reported that only 19% of students had not been taught the content by the time of the TIMSS assessment. In Belgium (Flemish), this content was intended to be taught to advanced students only, but teachers reported that just 22% of non-advanced students were not taught about relationships in simple food chains.

Additionally, while common sources of energy are not part of the national curricula in Finland, Poland, and Slovenia, only 25% of students in Finland, 2% in Poland, and 6% in Slovenia were reported by teachers as not being taught the topic. In contrast, the majority of students in these education systems received instruction on energy sources either in the assessed or preceding year, according to teacher reports.

4.3.2.2 Topics Taught and Student Achievement

The percentages of students who were taught (before and in grade four) and not taught were examined in relation to their mean achievement across the whole set of content areas and topics on environmental sustainability. Unexpectedly, the differences in mean achievement between the two groups were statistically significant in only a few cases across topics and education systems. For the topics of physical characteristics of Earth and Earth's rising average temperatures, the mean achievement of students not being taught these topics was lower than the achievement of students being taught these topics—though this pattern appeared in only 3 of the 28 education systems for each topic. In one education system (Kosovo), students not being taught these topics scored higher than students who were reported as having been taught them (Table 4.8).

The percentages of students not taught these topics were again lower than expected based on the reports on intended curricula. Although the topic of Earth's rising average temperatures was reported as not included in the national curricula of many education systems, teachers in systems such as Albania, Bulgaria, Finland, Hungary, Ireland, the Republic of North Macedonia and Spain, reported that the majority of students had been taught this content. Conversely, in Belgium (French), Cyprus, Montenegro, Poland, the Slovak Republic, and Slovenia, the majority of students were reported as not having been taught this content.

Similar declines in coverage of intended curricula were observed for the topic of physical characteristics of Earth, which was reported as not being included in the national curricula of education systems such as Albania, Finland, Ireland and Poland. However, the percentage of students not receiving instruction on this topic was relatively low in these systems. In Albania, 50% of students were not taught this content, whereas the figures were considerably lower in Finland (15%), Ireland (25%), and Poland (22%).

Most education systems reported high mean achievement among students learning life science and physical science topics, either in the assessment year or

Table 4.8 Teaching of earth science topics and students' environmental mean scale scores in education systems where difference is significant

Education system		Earth science: Physical characteristics of Earth			Earth science: Earth's rising average temperatures and results of this change		
		%	Mean	Sig.	%	Mean	Sig.
Belgium (Flemish)	Before	8.8 (2.5)	495.1 (6.5)		4.5 (1.5)	499.2 (12.1)	
	This year	23.0 (3.6)	502.8 (5.7)		50.3 (4.1)	494.9 (5.2)	
	Not yet	68.2 (3.8)	487.7 (4.0)	*	45.2 (4.2)	487.6 (5.2)	
Bulgaria	Before	34.2 (3.6)	565.3 (8.8)		15.4 (2.9)	552.6 (17.1)	
	This year	55.9 (3.5)	522.8 (9.7)		36.9 (4.1)	535.1 (11.2)	
	Not yet	9.9 (2.0)	491.5 (25.8)	*	47.7 (4.3)	529.8 (9.7)	
Hungary	Before	21.9 (3.5)	537.1 (10.8)		7.8 (1.8)	564.0 (8.3)	
	This year	60.1 (4.1)	534.2 (5.6)		48.4 (4.1)	535.8 (5.3)	
	Not yet	18.0 (3.3)	523.9 (9.3)		43.8 (4.0)	524.4 (7.4)	*
Kosovo	Before	22.5 (3.6)	404.8 (7.3)		12.5 (2.6)	392.2 (5.7)	
	This year	51.2 (4.3)	394.4 (6.0)		48.0 (4.4)	392.3 (6.5)	
	Not yet	26.3 (3.8)	415.9 (6.6)	*	39.6 (4.3)	417.4 (5.6)	*
Montenegro	Before	18.0 (2.4)	463.6 (6.3)		4.1 (0.8)	452.3 (11.1)	
	This year	55.7 (2.9)	466.9 (3.5)		21.5 (2.4)	479.7 (5.6)	
	Not yet	26.3 (2.9)	464.3 (5.1)		74.5 (2.5)	462.4 (2.6)	*
Slovak Republic	Before	31.3 (3.0)	522.4 (6.6)		11.1 (2.0)	531.8 (9.5)	
	This year	25.3 (2.9)	508.5 (9.6)		30.5 (2.6)	518.8 (7.0)	
	Not yet	43.3 (3.2)	499.6 (6.9)		58.3 (3.1)	499.6 (4.9)	*
Slovenia	Before	23.4 (2.8)	528.6 (5.0)		3.0 (1.1)	519.0 (10.8)	
	This year	47.8 (3.7)	517.9 (3.5)		28.6 (3.3)	524.5 (4.1)	
	Not yet	28.8 (3.2)	517.9 (4.2)	*	68.4 (3.5)	518.9 (3.3)	

Notes: Standard errors appear in parentheses. Due to rounding some values may seem inconsistent. The results for all education systems are in Table 4.17
*Significant difference between the mean scale scores at the time of teaching

prior. However, these differences were statistically significant in just a few cases. Education systems with the highest mean achievement for at least one topic—where students were taught the material either in the assessment year or in the previous year—include Montenegro, the Republic of North Macedonia, Portugal, and the Slovak Republic. It should be noted that students may also acquire knowledge about environmental sustainability outside of school settings.

4.3.2.3 Teachers Providing Opportunities to Learn Environmental Sustainability

Science teachers estimated how often they used four specific approaches when teaching environmental sustainability in their lessons: (1) develop students' positive attitudes toward the natural environment, (2) encourage students to use less resources, (3) discuss how student actions in and outside of school help the natural environment, and (4) discuss environmental issues. On average, teachers in all participating education systems reported using the four observed activities at least once a week or once or twice a month (see Table 4.18).

The frequency patterns of these approaches showed no clear association with the mean achievement of students. The most commonly practiced activity was developing students' positive attitudes toward the natural environment, reported as being practiced at least weekly in 20 out of 28 education systems. In half of the education systems, teachers also reported weekly encouragement for students to use fewer resources. Discussions about students' actions and environmental issues were less frequent, although these practices appear most directly related to teaching sustainability content.

Part of teaching sustainability is to engage students in specific activities. Education systems with high student achievement were more likely to include such engagement in activities or practice actions at the local level in their intended curricula, as indicated in the first part of the study. The proportions of students engaged in the four activities, as reported by teachers, varied significantly across education systems—from nearly all students to less than half (Fig. 4.3).

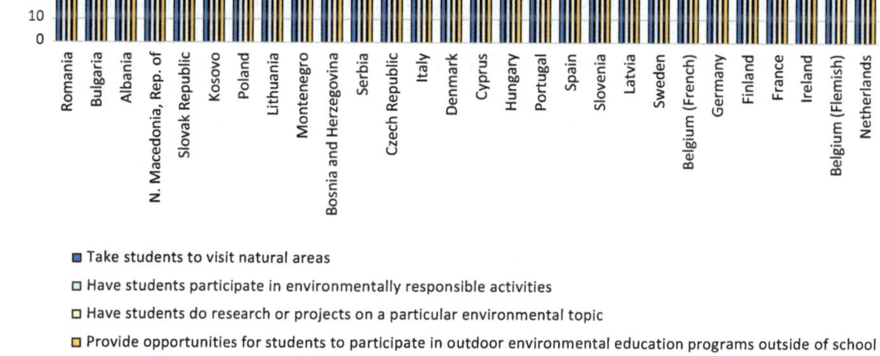

Fig. 4.3 Percentages of students whose teachers reported using different activities to teach about environmental issues and sustainability, by education system

In most education systems, the most common activity is involvement in environmentally responsible activities and the least common activity is providing opportunities to participate in outdoor environmental education programs outside of school. In Albania, Bulgaria, the Republic of North Macedonia, Romania, and the Slovak Republic, more than 70% of students were engaged in all four activities. In contrast, in France and the Netherlands, fewer than 50% of students participated in three of the four activities. It is worth noting that the percentages of students whose teachers reported engaging them in environmentally responsible activities is not aligned with intended curricula in all education systems. For example, Denmark and Ireland intend to include students into local environmental activities but their percentages of students engaged in those activities by their science teachers are not among the highest. The data clearly show the need for more information about who at the school is responsible for engaging students, and how schools engage them, in activities that most likely happen outside classes.

The significant differences in students' environmental knowledge between those who were or were not provided opportunities to participate in these activities are found in the following education systems (difference in number of points)[2]:

(a) Take students to visit natural areas: the Slovak Republic (16) and Germany (20).
(b) Have students participate in environmentally responsible activities: Spain (12), Denmark (19), Montenegro (21), Bosnia and Herzegovina (38), and Bulgaria (61).
(c) Have students do research or projects on a particular environmental topic (e.g., pollution, climate change): Denmark (11), the Czech Republic (12), Hungary (20), Kosovo (28), and Bulgaria (94).
(d) Provide opportunities for students to participate in outdoor environmental education programs outside of schools: Germany (19), the Republic of North Macedonia (29), Romania (36), and Albania (42).

In general, education systems where a positive relationship is observed between participation in activities and student achievement are primarily located in Eastern Europe and the Dinaric region. These include Albania, Bosnia and Herzegovina, Bulgaria, the Czech Republic, Hungary, Kosovo, Montenegro, the Republic of North Macedonia, Romania, and the Slovak Republic. In Western Europe, similar patterns are found in Denmark, Germany, and Spain.

An unexpected finding is the notably low percentage of students in Western Europe who are provided with opportunities to participate in outdoor environmental education programs outside of school. The data does not indicate whether such programs are unavailable outside of school or if teachers are simply not responsible for directing students to them. This underscores the need for further investigation into the opportunities available for students outside school to gain a fuller understanding of the situation.

---

[2] See Tables 4.12, 4.13, 4.14, and 4.15 for differences in achievement in environmental knowledge between students participating or not in environmental teaching activities provided by teachers for all education systems.

4.3.3 *Level of Attained Curricula*

4.3.3.1 Attitudes Toward Environment and Environmental Knowledge

The research question regarding attitudes was whether higher student knowledge is linked to more positive attitudes. Higher knowledge could lead to greater appreciation of nature or, conversely, more positive attitudes could motivate students to learn. However, the correlation analysis (Table 4.9) reveals that, in the observed education systems, environmental knowledge is less than moderately linked to student attitudes. Only a few correlations reached at least a moderate level (0.20 and higher). The notable exception is "Addressing climate change should be a high priority," which is moderately positively correlated with students' environmental knowledge in 11 education systems. In general, the strongest correlations were found in Denmark, followed by Belgium (Flemish) and Italy. Denmark, Italy, and Romania were the education systems that confirmed all the environmental content topics as being part of their national curricula. In contrast, Belgium (Flemish) reported that nearly all topics were taught only to the advanced students.

Correlations between achievement and agreement with the other three attitudinal statements are very weak or absent in the majority of education systems. Only in a few education systems is higher achievement linked to higher agreement, and this pattern differed across the three statements.

Nevertheless, the regression analysis shows that the Students Value Environmental Preservation scale scores—derived from their agreement to the statements in Table 4.9—generally predict environmental knowledge in all education systems, except four: Bosnia and Herzegovina, Poland, Serbia, and Slovenia (Table 4.10). The strongest association between valuing environmental preservation and

Table 4.9 The correlations between environmental knowledge and student attitudes toward environment

Addressing climate change should be a high priority								
Education system	Corr.	SE	Education system	Corr.	SE	Education system	Corr.	SE
Denmark	0.27	(0.03)	Ireland	0.23	(0.04)	Germany	0.20	(0.02)
Belgium (Flemish)	0.25	(0.02)	France	0.20	(0.02)	Spain	0.20	(0.02)
Italy	0.25	(0.04)	Bulgaria	0.20	(0.05)	Lithuania	0.20	(0.03)
Finland	0.24	(0.02)	Netherlands	0.20	(0.03)			

I care about the protection of plants and animals			It makes me sad when nature is destroyed			I enjoy being in nature		
Education system	Corr.	SE	Education system	Corr.	SE	Education system	Corr.	SE
Belgium (Flemish)	0.20	(0.03)	Italy	0.23	(0.03)	Sweden	0.24	(0.02)
Slovak Republic	0.20	(0.04)	Kosovo	0.21	(0.02)	Finland	0.20	(0.03)

Notes: SE, standard error. Correlations lower than 0.20 are not shown. Due to rounding some values may seem inconsistent

Table 4.10 Regression of association between Students Value Environmental Preservation scale and achievement in environmental knowledge

	Constant			Students Value Environmental Preservation						
Education system	B	se_B	t_B	B	se_B	b	b se	t_B	t_b	*p*
Bulgaria	414.3	36.7	11.3	11.0	0.2	3.1	0.1	3.6	3.7	$p < 0.05$
Albania	414.0	28.4	14.6	6.7	0.1	2.5	0.0	2.7	2.7	$p < 0.05$
Bosnia and Herzegovina	475.6	21.8	21.8	2.3	0.0	2.1	0.0	1.1	1.1	$p > 0.1$
Slovak Republic	437.4	20.0	21.9	7.5	0.1	1.9	0.0	3.9	4.3	$p < 0.05$
Italy	417.9	15.5	26.9	9.7	0.2	1.5	0.0	6.4	6.4	$p < 0.05$
Romania	487.1	16.9	28.9	3.7	0.1	1.5	0.0	2.5	2.5	$p < 0.05$
Denmark	459.4	15.0	30.6	8.5	0.2	1.4	0.0	5.9	6.1	$p < 0.05$
Ireland	449.9	15.9	28.2	8.4	0.2	1.4	0.0	6.2	6.3	$p < 0.05$
Lithuania	452,1	16.3	27.8	6.8	0.1	1.4	0.0	4.8	4.9	$p < 0.05$
Slovenia	500.6	14.0	35.8	2.1	0.0	1.4	0.0	1.4	1.4	$p > 0.1$
N. Macedonia, Rep. of	326.8	14.9	22.0	111	0.2	1.3	0.0	8.4	8.5	$p < 0.05$
Hungary	453.4	14.6	31.0	8.1	0.2	1.3	0.0	6.3	6.7	$p < 0.05$
Sweden	456.3	12.9	35.5	7.7	0.2	1.3	0.0	5.8	6.1	$p < 0.05$
Kosovo	332.5	13.9	23.9	6.9	0.2	1.3	0.0	5.4	5.9	$p < 0.05$
Portugal	468.7	13.4	35.0	4.7	0.1	1.3	0.0	3.5	3.5	$p < 0.05$
Serbia	497.3	13.3	37.5	2.2	0.0	1.3	0.0	1.6	1.6	$p > 0.1$
France	401.7	13.8	29.2	9.6	0.2	1.2	0.0	7.8	8.3	$p < 0.05$
Latvia	455.4	13.8	32.9	5.0	0.1	1.2	0.0	4.1	4.4	$p < 0.05$
Belgium (Flemish)	399.6	10.3	38.8	9.1	0.2	1.1	0.0	8.6	9.3	$p < 0.05$
Finland	471.8	12.0	39.3	7.5	0.2	1.1	0.0	6.9	7.4	$p < 0.05$
Cyprus	439.3	12.8	34.4	3.9	0.1	1.1	0.0	3.5	3.5	$p < 0.05$
Belgium (French)	392.4	11.1	35.5	9.0	0.2	1.0	0.0	8.6	8.7	$p < 0.05$
Germany	431.8	11.0	39.1	8.4	0.2	1.0	0.0	8.4	9.0	$p < 0.05$
Netherlands	462.9	8.7	53.5	5.9	0.2	0.9	0.0	6.8	7.0	$p < 0.05$
Poland	543.9	8.7	62.4	1.4	0.0	0.9	0.0	1.6	1.6	$p > 0.1$
Montenegro	404.1	9.4	43.1	6.3	0.1	0.8	0.0	7.7	7.6	$p < 0.05$
Czech Republic	476.7	7.5	63.8	4.5	0.1	0.8	0.0	5.7	5.7	$p < 0.05$
Spain	443.8	8.6	51.6	6.5	0.2	0.7	0.0	9.2	10.1	$p < 0.05$

environmental knowledge is observed in Bulgaria, followed by Albania and the Slovak Republic. The weakest association is seen in the Czech Republic, Montenegro and Spain.

Students were asked to report on their actual activities to help save the environment, including: reuse things and reducing resource usage (for example, water, food); discussing ways to help the environment (for example, saving water, picking up trash); learning about environmental problems (for example, climate change, endangered species); participating in group activities to help the environment; and, informing friends when they are doing things that harm the environment.

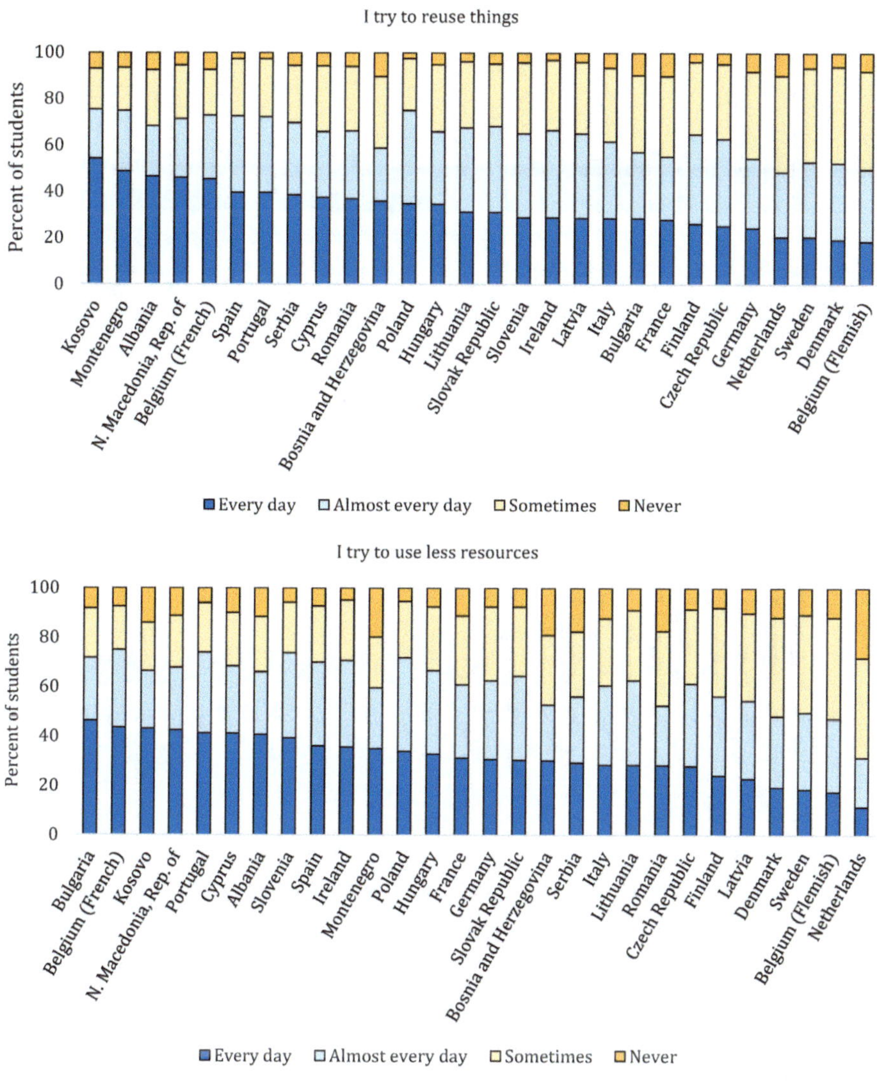

Fig. 4.4 Frequency of student actions: reuse and use less resources

All these activities require a specific level of knowledge to be performed, therefore they are regarded as part of the attained curriculum.

In general, students reported engaging in these activities relatively often (Figs. 4.4, 4.5, and 4.6). The percentage of students who never participated in the observed activities is low in most education systems. The most common activity reported is reusing things, while the least common is participating in group activities to help the environment.

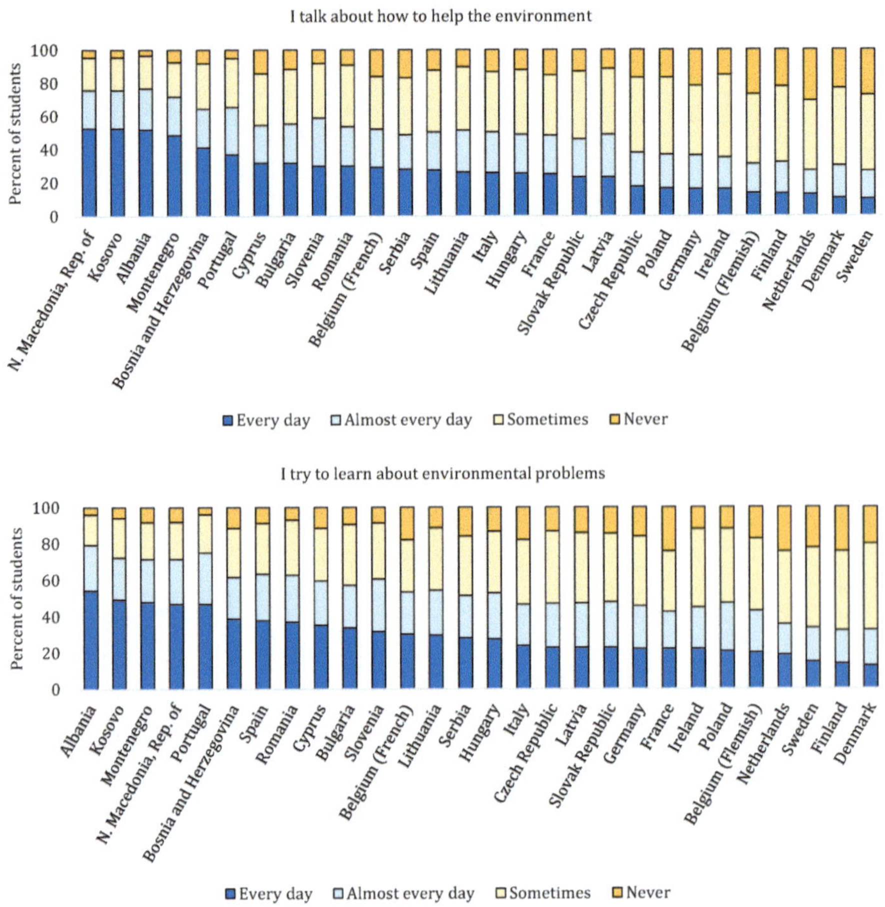

Fig. 4.5 Frequency of student actions: talk and learn about environment

Reusing items is most frequently reported by students in Kosovo, Montenegro, and Albania, while students in Bulgaria, Belgium (French), and Kosovo are the most likely to try to use fewer resources, with around 70% of students engaging in these behaviors every day or almost every day. On the other hand, students in Belgium (Flemish), Denmark, the Netherlands, and Sweden—education systems within the European Union—report the lowest rates of reusing and using less resources.

Students report trying to learn about the environment and discuss caring for it most often in Balkan education systems and least often in Scandinavian and Western European education systems including Denmark, Finland and the Netherlands (Fig. 4.6). Students are most likely to participate in group activities and warn their friends about environmental mistakes in Albania, Bosnia and Herzegovina, Kosovo, the Republic of North Macedonia, and Montenegro with more than half of students

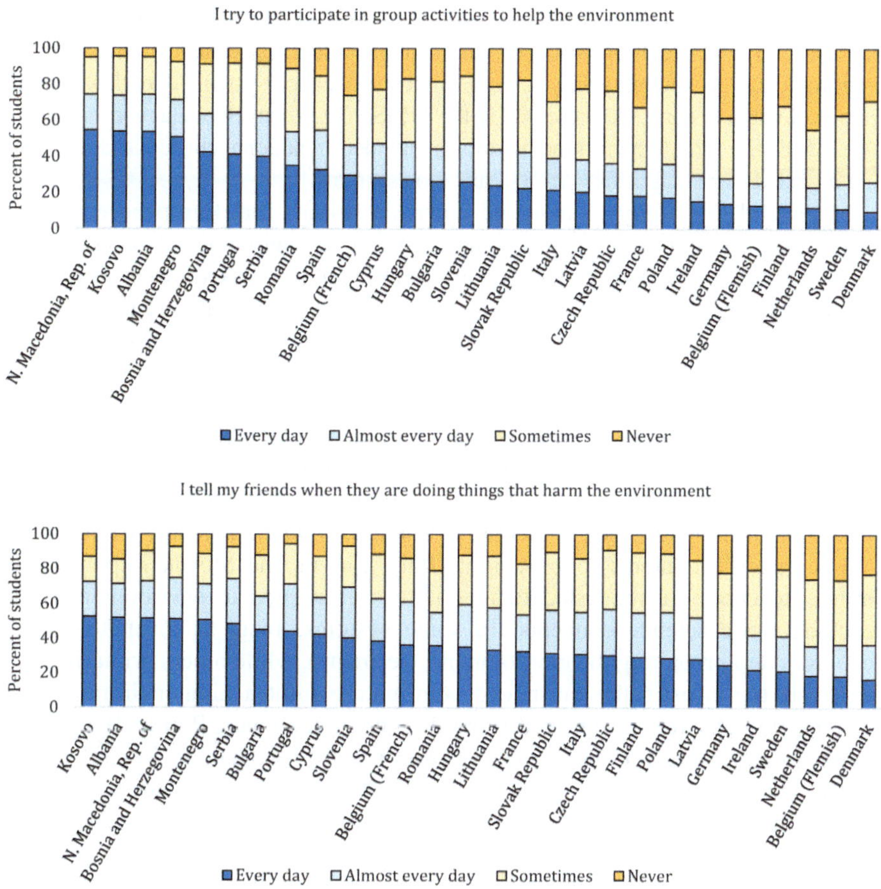

Fig. 4.6 Frequency of student actions: activities in groups and warning friends

engaging in these behaviors every day and an additional 20% almost every day. In contrast, education systems like Belgium (Flemish), Denmark, Finland, the Netherlands, and Sweden report that less than a third of students participate in these activities every day or almost every day (Fig. 4.6).

4.4 Discussion

This study of intended, implemented, and attained curriculum in the area of environmental knowledge aimed to address four key research questions.

The findings provide an affirmative answer to the first research question—whether national curricula, teaching practices, and student outcomes related to education for environmental sustainability vary across European education systems. A review of summaries of national curricula presented in the TIMSS 2023 encyclopedia reveals that the extent and structure of environmental topics included in national curricular documents and intended to be taught vary across education systems.

In general, the study of intended curricula reveals that European education systems do not uniformly cover essential aspects of environmental sustainability. Topics related to living nature are better covered than physical sciences. For example, the sustainable use of energy and climate change are not taught in a third to more than half of the education systems. However, the positive and negative impacts of humans on the environment are taught across all of Europe in every observed education system in this study.

There are differences in the contents requested to be taught by grade four and approaches to teaching environmental sustainability. Some education systems focused their curriculum on developing positive attitudes, values and behaviors toward environmental issues (Belgium, Finland, Ireland, Poland), while others outline specific contents to be taught (Italy, Latvia, the Slovak Republic). In France, teachers are requested to identify issues related to environment when teaching basic science concepts. In some systems, focus is given to use of resources and protecting the environment (Germany), while others primarily support fostering investigative attitudes toward nature and teaching environmentally responsible behaviors (the Czech Republic, the Netherlands, Portugal) or developing responsibility for environmental sustainability (Romania). Several education systems also explicitly require students to develop an understanding of human interaction with nature (Bulgaria, Denmark, Montenegro, Portugal, Sweden) with a particular focus on pollution, recycling and waste management to preserve nature (Serbia, Slovenia). Spain, by contrast, emphasizes teaching general interactions between living things.

Education systems with the highest mean achievement differ from others in that their curricula explicitly encourage student engagement or practice participation in activities of caring for the environment on local levels. However, the available curricular descriptions are general and are open to interpretation regarding how content is taught in schools. It is likely that these curricula are supported by additional teaching materials and instructions which were not observed in this study.

One of the aims of this study was to identify exemplary practices across Europe. Finland, Ireland, and Poland stand out as education systems with clearly defined and explicit environmental education goals within their curricula. For example, Finland's curriculum states:

> Building a Sustainable Future: Students study biodiversity, climate change, and its mitigation; sustainable use of natural resources; promoting health; caring for one's own

> cultural heritage; living in a multicultural world; and the global wellbeing of humanity now and in the future. Students take part in a collaborative project in which they practice participation and involvement at the local or global level.

Poland and Ireland directly request teaching of environmental awareness and to develop values. Poland emphasizes:

> ... learning the values and interrelationships of components of the natural environment, learning the values and norms that source a healthy ecosystem, and developing behaviours resulting from these values.

Ireland specifies in the main content of the *Environmental Awareness* unit:

> ...identify positive aspects of natural and built environments; identify the interrelationship of living and nonliving elements of local and other environments; become aware of the importance of Earth's renewable and non-renewable resources; recognize how people's actions affect their environment; come to appreciate the need to conserve resources; ... identify some ways in which science and technology contribute positively to society; recognize and investigate human activities that have positive and negative effects on local and wider environments ... caring for the Environment, examine a number of ways in which the local environment could be improved or enhanced; identify and discuss a local, national, or global environmental issue; realize that there is a personal and community responsibility for taking care of the environment.

Ireland's curriculum statements go beyond the list of observed topics in this study, emphasizing activities and higher-order cognitive skills encouraging students to appreciate, investigate, problem-solve, discuss, and take responsibility for environmental care. The structure aligns with key competencies such as critical thinking, community participation, and interdisciplinary connections—competencies widely recognized as essential for educators to address sustainability challenges (Corres et al., 2020).

Given their explicit curricular emphasis and high student achievement, these education systems provide positive examples of intended curricula for teaching environmental sustainability.

Variations in how environmental sustainability content is taught across education systems is also evident in their reports on topics assessed in TIMSS. While most education systems confirmed that these topics are taught to all students—consistent with the TIMSS aim of assessing knowledge acquired in school—some systems did not cover certain topics. Two topics were omitted in the majority of education systems: fossils and competition in ecosystems, which are most likely addressed in higher grades than grade four. Moreover, the issue of Earth's rising average temperatures and climate change was reported as not intended to be taught to any students in more than half of the 28 education systems and in 4 systems was reserved for advanced students, despite many systems acknowledging environmental issues in their curricula.

Overall, the analysis of inclusion of specific topics across curricula reveals that, when asked about the teaching of specific topics, education systems report a higher degree of intended coverage than might be inferred from the general curriculum documents. This finding underscores the need for further investigation into the curricular foundations of student knowledge.

A review of the linkage between Sustainable Development Goals and education outcomes (Kioupi & Voulvoulis, 2019) found that the most relevant pedagogies for fostering sustainability competencies engage the cognitive (head), psychomotor (hands), and affective (heart) domains. While all three domains are present across the observed education systems, all three are not present in each. Effective strategies include problem-based learning, active learning, community service learning, critical emancipatory pedagogy, place-based environmental education, ecojustice education, community learning, and traditional ecological knowledge. However, evidence from TIMSS show that many of these elements are already integrated into European curricula.

The examination of implemented curricula and the relationship with intended curricula topics, topics taught by teachers, and student achievement provides positive findings. The answer to the second research question—how knowledge is linked to content taught in schools—is not straightforward. On average, the achievement of students who had not yet been taught specific content was not lower than the achievement of students who had been taught the topics before taking the assessment. This suggests that students also acquire knowledge about environmental topics outside the classroom. Only in some education systems, and only for some topics, was the achievement of students who had not been taught the content lower than the achievement of students who had already been introduced to these concepts.

A positive finding of the analysis is that teachers across the European education systems reported teaching more topics than were explicitly outlined in curricular documents. There are only a few cases of education systems where notable percentages of students were reported not to have been taught some contents which are part of the prescribed curricula. It should also be noted that TIMSS assessments were administered several months before the end of school year, leaving teachers time to teach some concepts after reporting them as not yet taught.

The findings underscore the importance of involvement of teachers in teaching the content they believe students need to be taught. While it is important that teachers are well-informed about the content they are expected to teach, they also need support and autonomy to quickly adapt their instruction to meet students' growing learnings needs, including areas beyond the scope of the national curriculum.

Across Europe, there are significant differences in implemented teaching content. In some education systems, the curriculum is demanding, yet item coverage reported after assessments is low (for example, in Belgium [Flemish]). A comparative study of curricula from four Eastern European education systems—the Czech Republic, Estonia, Poland, and Slovenia—highlighted substantial differences in the breadth of learning outcomes and topics (Kácovský et al., 2021). For example, Slovenia's highly detailed curriculum contains six times as many learning outcomes as the much briefer Czech curriculum for the same science subjects; however, higher-order cognitive goals remain largely absent from Slovenia's curriculum compared to other examined curricula.

The main findings regarding attained curricula indicate differences in student achievement in environmental knowledge that do not reveal systematic patterns according to geographic or other characteristics of the observed education systems.

Within attained curricula, students' attitudes toward nature are also considered, as these are assumed to develop alongside learning about environmental topics. Findings about the relationship between environmental awareness and achievement highlight the importance of teaching environmental content. The weak or nonexistent correlations between students' agreement with statements about enjoying nature and their achievement suggest that students appreciate nature regardless of their knowledge or learning opportunities. A more systematic pattern emerges in students' agreement with the statement, "Addressing climate change should be a high priority." While the relationship between this belief and achievement remains weak, it is consistently observed in most education systems. Unlike other statements related to enjoying nature, this one requires students to understand the concept of climate change and the meaning of priority—both of which must be explicitly taught and can explain stronger links to achievement than agreements with other statements. The strongest relationship is found in Belgium (Flemish), Denmark, Finland, and Italy. In Finland, climate change is part of the curriculum and in the other three education systems the curriculum requires the development of students' understanding of environmental problems and of the interaction between humans and nature. This suggests that higher levels of environmental knowledge may strengthen the link between attitudes and knowledge, particularly when some conceptual understanding is needed to form an opinion about the issue.

Furthermore, the Students Value Environmental Preservation scale, based on students' agreement with environmental statements, was found to predict higher environmental knowledge in all education systems except four. The strongest associations were observed in Albania, Bulgaria, and the Slovak Republic—systems that differ substantially in their mean achievement levels but are all located in Eastern Europe. In general, the development of positive attitudes toward nature appears to play an important role in enhancing students' knowledge.

Unexpected patterns emerge in student reports of their actual engagement in environmental activities requiring some knowledge. The highest percentages of students—over 70%—who actively try to save resources, learn about the environment, and participate in environmental activities daily or almost daily are found in Balkan and Eastern European education systems, including some that are not European Union members. These systems may be particularly active in environmental initiatives as part of their efforts to join the European Union, frequently participating in large-scale projects with other European nations. For example, a study comparing four European education systems (Kovách et al., 2021) found that in Serbia, many schools take part in the Eco-Schools program, a global initiative involving 56,000 schools. Additionally, many Eastern European and Balkan schools engage in green projects run by nongovernmental organizations or under the Erasmus program such as the Education for Climate Community.[3]

In contrast, Scandinavian education systems report significantly lower student participation in environmental activities, with fewer than a third of students engaging in them daily or almost daily. While the present study does not provide explanations for the lower involvement of younger students in Scandinavian education systems, it highlights the need for further research.

Finally, regarding the growing trend of gender differences in science achievement and attitudes in TIMSS, the majority of results across the education systems in this study did not differ by gender. This finding aligns with previous research, such as the study by Liefländer and Bogner (2014), which examined gender and age differences in environmental awareness. Their study concluded that environmental education is more effective in fostering pro-environmental attitudes in younger students (aged 9–10) than in older students, and that gender does not necessarily influence the development of such attitudes.

4.5 Limitations of the Study

This study aimed to address gaps in research regarding the relationships between teaching, knowledge, and the development of attitudes toward environmental awareness, particularly among young children and through cross-national comparisons of education systems. While the study utilized multiple sources of data, each source has inherent limitations, particularly when combined. One limitation is that the TIMSS study does not provide data for all European Union member education systems or for all non-European Union education systems. The findings of this study are limited to the education systems participating in TIMSS with their grade four student populations. The primary source, the curricula summaries in the TIMSS 2023 encyclopedia (Reynolds et al., 2024), are not in a comparable form. The curricular documents have been, for most systems, summarized into general overviews, which can lead to an underestimation of prescribed content. Some systems may include more environmental topics in teaching through national policies, such as school rules for saving resources or separating waste. Then, some teachers may underreport the teaching of a topic if they have not completed it. Future studies with a specific focus on environmental contents are needed to reveal the full extent of differences across Europe. Finally, the indices and sets of questions on sustainability used in this study were developed a few years before the data collection in 2023. As environmental

[3] Accessible at: https://education-for-climate.ec.europa.eu/community/

sustainability is a rapidly evolving issue, the relevance of some concepts may have shifted. To accurately capture the current situation in Europe, future studies should develop new or adapted questions, statements, and variables to capture emerging issues.

Unexpected findings also emerged regarding student behaviors, such as students reusing materials. Interestingly, some economically strong education systems exhibited lower rates of reuse among students. While economic factors may partially explain this, the available data do not allow for a detailed analysis or overview of the extent of recycling across the education systems, highlighting the need for future research to more comprehensively address the issue.

4.6 Conclusion

The TIMSS data on education for environmental sustainability, which includes information on environmental knowledge, attitudes, student activities, and content taught, provides a valuable, though complex, source for comparing education systems across three levels of curricula—intended, implemented, and attained—revealing important insights.

In summary, national curricula, teaching approaches, and student outcomes in education for environmental sustainability vary across European education systems. The national curricula show large differences in the presentation and scope of content intended to be taught, including topics and expected learning outcomes, ranging from direct content knowledge of environmental issues to engagement of students in activities and practical actions to care for nature. Teachers report different levels of coverage of the topics from their national curricular documents; in general, they frequently teach more than is indicated in official curricula. Student outcomes vary across Europe, although these differences do not follow clear geographic patterns.

The comparison of national curricula and mean achievement reveals that education systems emphasizing understanding of environmental issues and student engagement in practical activities at the local level, tend to show higher student knowledge. An unexpected finding is that students who had not yet been taught specific content demonstrated knowledge levels comparable to students who had been taught these in science classes. This indicates that environmental concepts are also learned outside science classes—reporting of content taught in science may not fully capture learning that occurs elsewhere, even in school.

Another unexpected result was that higher student knowledge is generally not directly linked to basic positive attitudes like enjoyment of nature; low achieving students also expressed enjoyment of nature and care for living beings. Higher achievement is, however, linked to general environmental awareness, complex attitudinal scales, and attitudes towards issues that require understanding and knowledge, such as climate change. Therefore, the answer to the third research question—whether higher student environmental knowledge is linked to more positive attitudes toward environmental sustainability—is affirmative: more positive complex attitudes and opinions about specific issues are linked to higher knowledge in a large number of education systems.

In relation to the fourth and final research question—addressing whether groups of European education systems share similarities in teaching approaches, student knowledge, or attitudes toward environmental sustainability—no clear geographical pattern in environmental knowledge and national curricula was identified. However, greater student engagement in environmental activities was observed in the eastern part of Europe than in the western part.

The teaching of environmental content across Europe varies; however, some education systems may serve as models for others in developing their goals and curricular documents, with respect to particular aspects of environmental education. Some show strengths in national curricular documents, others in engagement of students in environmental activities. The findings suggest that more frequent teaching and learning activities lead to greater knowledge, stronger connections between attitudes, and better preparedness among young students for environmental preservation activities. However, certain content areas—such as green energy, renewable resources, and climate change, which are widely discussed in public media—are still not systematically included in the curriculum of most education systems. The available sources also lacked data which would allow for greater understanding of the existing reports about intended curricula and, in particular, the rationale for including specific content, practical activities implemented in schools, and the responsibilities of teachers and schools to provide and engage students in activities outside schools. More precise national descriptions would support future research on the inclusion of students in environmental actions.

Appendix (Tables 4.11, 4.12, 4.13, 4.14, 4.15, 4.16, 4.17, and 4.18)

Table 4.11 Coverage of items by curricula in each education system

	ITA	MKD	ROM	LTU	SWE	BIH	DNK	DEU	PRT	SRB	IRL	MNE	CZE	FRA	ESP	ALB	LVA	NLD	BFR	FIN	SVK	HUN	XKX	SVN	POL	BFL	Total
SE71069	1	1	1	1	1	1	1		1	1	1	1	1	1	1	1	1	1	1	1	1	1	1		1		23
SE71080	1	1	1	1	1	1	1	1	1	1	1		1		1	1	1	1	1			1	1	1	1		21
SE71220	1	1	1	1	1	1	1	1						1								1					10
SE61015	1	1	1	1	1	1	1	1	1	1	1	1	1	1	1	1	1	1	1	1	1	1	1	1	1	1	26
SE61124	1	1	1	1	1	1	1	1	1	1	1	1	1	1		1			1	1	1	1					19
SE61116	1	1	1	1	1	1	1	1	1	1	1	1	1			1	1	1		1	1					1	19
SE81083	1	1	1	1	1	1	1	1	1	1	1	1	1	1	1		1	1	1	1		1	1		1		22
SE81073	1	1	1	1	1	1	1	1	1	1	1	1	1	1	1	1	1	1	1	1	1		1		1		23
SE81054	1	1	1	1	1	1	1	1	1	1	1	1			1	1	1				1	1	1	1	1		20
SE81036	1	1	1	1	1	1	1	1	1	1	1	1	1	1	1	1	1	1	1	1	1	1	1	1			24
SE81075	1	1	1	1	1	1	1	1	1	1	1	1	1	1	1		1	1	1		1				1		20
SE81920A	1	1	1	1	1	1	1	1	1	1	1	1	1	1	1	1			1	1	1	1	1	1	1	1	24
SE81920B	1	1	1	1	1	1	1	1	1	1	1	1	1	1	1	1			1	1	1	1	1	1	1	1	24
SE81204	1	1	1	1	1	1	1	1	1	1	1	1	1	1	1				1		1	1	1	1			20
SE81070	1	1	1	1	1	1	1	1	1	1	1	1	1	1	1		1		1				1	1			19
SE81055	1	1	1	1	1	1	1	1	1	1	1	1	1	1	1	1	1	1		1	1	1	1	1		1	24
SE81207A	1	1	1	1	1	1		1	1	1	1	1		1	1	1	1		1	1	1						18
SE81207B	1	1	1	1	1	1		1	1	1	1	1		1	1	1	1		1	1	1						18
SE71063	1	1	1	1	1	1	1	1	1	1	1	1	1	1	1		1	1	1				1	1			20
SE71214	1	1	1	1	1	1	1	1	1	1	1	1	1	1	1	1	1	1	1	1	1	1		1	1		24
SE71213	1	1	1	1	1	1	1	1	1	1	1	1		1		1			1		1	1					17
SE81044	1	1	1	1	1	1	1	1	1	1	1	1	1	1	1	1	1	1	1	1	1	1		1		1	24
SE81042	1	1	1	1	1	1	1	1	1	1			1	1	1	1	1	1						1	1		18

(continued)

Table 4.11 (continued)

	ITA	MKD	ROM	LTU	SWE	BIH	DNK	DEU	PRT	SRB	IRL	MNE	CZE	FRA	ESP	ALB	LVA	NLD	BFR	FIN	SVK	HUN	XKX	SVN	POL	BFL	Total
SE81202	1	1	1	1	1	1	1	1	1	1	1	1	1	1	1	1	1	1	1	1	1	1	1	1			24
SE81239	1	1	1	1	1	1	1	1	1	1	1		1		1	1	1	1		1	1			1			19
SE71076	1	1	1	1	1	1	1	1	1	1	1	1	1	1	1	1	1	1	1	1		1	1		1		23
SE71223	1	1	1	1	1	1	1	1	1	1	1	1	1	1	1		1	1	1	1	1	1	1	1	1	1	25
SE81080	1	1	1	1	1	1	1	1	1	1	1	1	1		1	1	1	1	1				1	1			20
SE81168	1	1	1	1	1	1	1	1	1	1	1	1	1	1	1	1	1	1	1	1	1	1	1	1	1	1	26
SE81085	1	1	1	1	1	1	1	1	1	1	1	1	1	1	1	1	1	1	1	1	1		1		1	1	24
SE81238	1	1	1	1	1		1		1	1	1	1	1	1		1	1		1	1	1	1		1		1	20
SE81200	1	1	1	1	1	1	1	1	1	1		1	1	1	1	1	1	1				1	1	1	1		21
SE81079	1	1	1	1	1	1	1	1	1	1	1	1	1	1	1	1	1	1	1	1	1		1				22
SE81076	1	1	1	1	1	1	1	1	1	1	1	1	1	1	1	1	1	1		1	1	1	1		1	1	24
SE81077	1	1	1	1	1	1	1	1	1	1	1	1	1	1	1	1	1	1	1	1	1	1	1	1	1	1	26
SE81212	1	1	1	1	1	1	1	1	1	1	1	1		1		1		1		1	1	1	1				19
SE61069	1	1	1	1	1	1	1	1	1	1	1	1	1	1	1	1	1	1		1			1	1	1		22
SE61019	1	1	1	1	1	1	1	1	1	1	1	1	1	1	1	1	1	1	1	1	1	1		1	1		24
SE71071	1	1	1	1	1	1	1	1	1	1	1	1	1	1	1	1	1	1	1	1	1	1		1	1		24
SE71222	1	1	1			1	1	1				1		1								1					9
SE71065	1	1	1	1	1	1	1	1	1	1	1	1	1	1	1	1	1	1	1	1			1		1		22
SE71081	1	1	1	1	1	1	1	1	1	1	1	1	1		1	1	1	1	1				1				19
SE71201	1	1	1	1	1	1	1	1	1	1	1	1	1	1	1	1	1	1		1	1	1	1	1			23
SE71237	1	1	1	1	1		1	1	1	1	1	1	1		1	1	1	1		1	1			1	1		20

Note: 1, the item is covered by the education system; ALB, Albania; BFL, Belgium (Flemish); BFR, Belgium (French); BIH, Bosnia & Herzegovina; CZE, Czech Republic; DEU, Germany; DNK, Denmark; ESP, Spain; FIN, Finland; FRA, France; HUN, Hungary; IRL, Ireland; ITA, Italy; LTU, Lithuania; LVA, Latvia; MKD, North Macedonia; MNE, Montenegro; NLD, Netherlands; POL, Poland; PRT, Portugal; ROM, Romania; SRB, Serbia; SVK, Slovak Republic; SVN, Slovenia; SWE, Sweden; XKX, Kosovo

Table 4.12 Differences in achievement in environmental knowledge between students participating or not in environmental teaching activities by teachers: visiting natural areas

	Take students to visit natural areas (e.g., a pond or meadow)						
Education system	Yes (% of students)	SE	Mean	SE	Diff. (yes/no)	SE	Sig.
Finland	97.30	(0.97)	543.17	(3.35)	11.08	(12.54)	
Albania	95.13	(1.73)	478.29	(6.41)	13.67	(18.77)	
Lithuania	91.66	(2.36)	519.73	(3.55)	8.93	(12.04)	
Bosnia and Herzegovina	90.68	(2.36)	447.18	(5.21)	15.88	(10.58)	
Latvia	90.18	(1.98)	503.63	(3.65)	1.53	(8.74)	
Slovak Republic	87.32	(2.44)	506.89	(4.64)	15.88	(7.44)	*
Romania	86.97	(3.04)	522.34	(6.39)	23.49	(16.28)	
Slovenia	86.97	(2.52)	519.43	(3.13)	4.47	(4.60)	
Belgium (Flemish)	86.61	(3.05)	490.61	(3.64)	7.23	(7.15)	
Montenegro	86.34	(2.30)	465.05	(2.73)	3.95	(6.66)	
N. Macedonia, Rep. of	84.17	(2.88)	437.38	(6.03)	3.93	(14.24)	
Czech Republic	82.30	(2.72)	520.20	(3.48)	1.29	(7.66)	
Serbia	81.70	(3.04)	517.74	(3.55)	3.59	(8.21)	
Germany	81.14	(2.76)	516.69	(3.31)	20.35	(10.12)	*
Denmark	80.45	(3.50)	534.61	(4.02)	4.40	(7.73)	
Kosovo	76.83	(3.67)	400.48	(5.05)	8.83	(9.32)	
Spain	75.78	(2.13)	510.54	(2.88)	1.79	(5.69)	
Poland	75.17	(3.38)	556.68	(2.54)	1.16	(5.66)	
Bulgaria	73.20	(3.09)	526.74	(7.50)	21.74	(12.29)	
Hungary	71.13	(4.06)	536.57	(5.13)	12.37	(10.95)	
Sweden	70.54	(3.75)	528.71	(4.96)	3.74	(7.58)	
Italy	67.87	(3.06)	519.68	(3.39)	7.21	(5.44)	
Cyprus	63.27	(3.41)	481.29	(3.99)	5.02	(6.23)	
Portugal	60.13	(3.14)	520.06	(4.23)	4.08	(5.52)	
Belgium (French)	57.46	(3.41)	485.03	(4.25)	5.70	(6.43)	
France	46.48	(4.47)	498.43	(6.02)	5.09	(7.00)	
Ireland	46.46	(4.25)	540.01	(5.19)	6.92	(6.78)	
Netherlands	39.46	(5.32)	525.40	(4.50)	10.71	(5.74)	

Note: SE, standard error

*Significant difference between the mean scale scores at the time of teaching

Table 4.13 Differences in achievement in environmental knowledge between students participating or not in environmental teaching activities by teachers: participating in environmentally responsible activities

Education system	Have students participate in environmentally responsible activities (e.g., pick up trash)						
	Yes (% of students)	SE	Mean	SE	Diff. (yes/no)	SE	Sig.
Bulgaria	98.94	(0.78)	531.93	(6.01)	60.61	(11.88)	*
Romania	98.74	(1.25)	525.06	(6.04)	–	–	
Kosovo	98.67	(0.98)	402.97	(4.07)	48.38	(65.76)	
Slovak Republic	97.73	(0.94)	508.22	(4.47)	29.93	(15.59)	
Hungary	97.54	(1.08)	533.41	(4.32)	16.74	(26.13)	
Albania	96.07	(2.88)	478.63	(6.38)	8.26	(16.03)	
Montenegro	94.09	(0.94)	467.08	(2.81)	21.50	(8.30)	*
Bosnia and Herzegovina	93.65	(2.61)	446.26	(5.28)	37.85	(9.33)	*
Lithuania	92.91	(1.70)	519.81	(3.27)	8.50	(13.78)	
Serbia	92.87	(1.83)	518.67	(3.38)	22.27	(13.22)	
Slovenia	92.32	(1.72)	519.54	(3.07)	7.29	(6.11)	
N. Macedonia, Rep. of	90.40	(2.62)	438.54	(5.56)	18.63	(22.39)	
Finland	86.74	(2.10)	543.13	(3.35)	1.94	(5.97)	
Belgium (Flemish)	85.47	(3.18)	492.44	(3.88)	5.98	(12.44)	
Belgium (French)	83.32	(2.42)	483.48	(3.40)	4.63	(9.63)	
Poland	83.08	(2.76)	555.45	(2.69)	8.95	(6.29)	
Latvia	81.70	(3.39)	504.45	(3.89)	2.45	(7.46)	
Czech Republic	81.63	(2.76)	520.25	(3.64)	1.95	(6.00)	
Italy	81.13	(2.60)	517.47	(2.93)	0.56	(6.75)	
Portugal	80.78	(2.63)	517.83	(3.54)	3.16	(6.71)	
Spain	79.91	(2.26)	507.77	(2.91)	11.63	(5.20)	*
Denmark	79.74	(3.38)	539.34	(3.94)	19.21	(6.26)	*
Germany	78.59	(2.84)	514.79	(3.63)	9.08	(8.79)	
Ireland	76.37	(3.46)	534.81	(4.27)	6.31	(8.66)	
France	74.44	(3.05)	495.21	(4.77)	2.26	(8.22)	
Cyprus	72.27	(3.04)	480.27	(3.68)	2.99	(7.46)	
Sweden	71.58	(3.65)	527.59	(5.01)	0.04	(6.82)	
Netherlands	61.07	(5.25)	520.00	(4.39)	2.80	(7.27)	

Note: SE, standard error; –, result not reported because estimation is not reliable
*Significant difference between the mean scale scores at the time of teaching

Table 4.14 Differences in achievement in environmental knowledge between students participating or not in environmental teaching activities by teachers: doing research or projects

	Have students do research or projects on a particular environmental topic (e.g., pollution, climate change)						
Education system	Yes (% of students)	SE	Mean	SE	Diff. (yes/no)	SE	Sig.
Romania	96.94	(1.59)	525.68	(6.05)	9.21	(7.71)	
Albania	94.94	(2.69)	480.46	(6.47)	29.81	(16.06)	
N. Macedonia, Rep. of	92.15	(2.15)	437.81	(5.71)	13.39	(23.64)	
Portugal	92.10	(1.93)	518.74	(3.66)	3.85	(13.92)	
Kosovo	91.43	(2.61)	404.71	(4.26)	27.63	(12.40)	*
Serbia	89.59	(2.26)	517.17	(3.33)	0.86	(10.71)	
Bulgaria	86.92	(2.42)	544.89	(5.75)	94.23	(23.70)	*
Italy	83.96	(2.47)	517.20	(2.89)	2.71	(7.63)	
Spain	76.37	(2.40)	509.75	(3.17)	0.73	(5.11)	
Bosnia and Herzegovina	74.48	(4.04)	449.15	(6.17)	1.90	(8.91)	
Ireland	73.49	(3.13)	537.45	(4.42)	4.31	(5.80)	
Slovak Republic	72.50	(2.92)	508.58	(5.41)	1.18	(8.77)	
Montenegro	72.36	(2.41)	466.98	(2.98)	4.46	(6.87)	
Lithuania	67.10	(3.71)	518.59	(3.99)	5.72	(6.20)	
Germany	64.42	(3.49)	517.59	(4.11)	13.27	(6.85)	
Sweden	62.99	(4.42)	527.06	(5.43)	1.47	(7.73)	
Denmark	61.34	(4.34)	539.79	(3.76)	11.15	(5.16)	*
Slovenia	61.34	(3.22)	519.31	(3.70)	1.82	(4.27)	
Finland	57.33	(3.20)	545.43	(4.11)	5.98	(5.10)	
Poland	57.11	(3.38)	557.11	(3.66)	0.33	(4.41)	
Cyprus	54.29	(3.90)	481.03	(4.24)	3.47	(6.21)	
Latvia	54.16	(3.88)	504.61	(4.97)	1.81	(6.47)	
Hungary	50.63	(3.98)	542.99	(6.06)	20.23	(8.95)	*
France	44.49	(4.27)	500.66	(5.56)	9.47	(6.95)	
Belgium (Flemish)	43.85	(3.83)	496.27	(5.57)	8.37	(5.95)	
Czech Republic	42.52	(3.04)	526.84	(4.56)	11.95	(4.33)	*
Belgium (French)	39.10	(3.47)	480.57	(5.71)	3.78	(7.15)	
Netherlands	37.90	(4.65)	523.67	(5.29)	7.67	(6.36)	

Note: SE, standard error
*Significant difference between the mean scale scores at the time of teaching

Table 4.15 Differences in achievement in environmental knowledge between students participating or not in environmental teaching activities by teachers: participating in programs outside of school

Education system	Provide opportunities for students to participate in outdoor environmental education programs outside of school						
	Yes (% of students)	SE	Mean	SE	Diff. (yes/no)	SE	Sig.
Romania	9.70	(2.36)	523.49	(6.27)	36.08	(12.92)	*
Bulgaria	78.22	(3.39)	540.43	(6.47)	31.62	(23.73)	
Albania	76.45	(5.35)	488.90	(6.64)	42.26	(18.45)	*
N. Macedonia, Rep. of	74.87	(3.28)	429.45	(6.03)	29.09	(12.45)	*
Slovak Republic	72.23	(3.24)	507.78	(5.08)	4.04	(7.15)	
Kosovo	69.53	(4.00)	400.57	(5.21)	5.69	(8.66)	
Poland	69.38	(3.80)	556.07	(2.77)	2.93	(5.62)	
Lithuania	69.27	(3.47)	519.72	(3.85)	2.44	(6.97)	
Montenegro	67.86	(2.15)	468.35	(3.05)	8.59	(5.30)	
Bosnia and Herzegovina	63.37	(5.62)	455.14	(5.24)	17.68	(8.92)	
Serbia	62.94	(3.55)	516.79	(3.98)	0.79	(6.29)	
Czech Republic	60.83	(3.25)	521.39	(3.99)	3.63	(4.49)	
Italy	60.41	(3.27)	517.70	(2.95)	0.86	(5.67)	
Denmark	59.35	(4.30)	537.71	(4.28)	5.50	(5.70)	
Cyprus	56.41	(3.48)	476.36	(4.42)	7.07	(7.01)	
Hungary	55.28	(4.40)	539.02	(5.54)	13.46	(9.52)	
Portugal	54.11	(3.37)	519.10	(3.79)	1.45	(5.34)	
Spain	51.71	(2.70)	508.77	(3.56)	2.50	(4.57)	
Slovenia	51.01	(3.58)	519.31	(4.06)	1.45	(3.93)	
Latvia	47.63	(4.29)	505.24	(5.18)	2.80	(6.78)	
Sweden	41.16	(4.44)	529.58	(4.87)	3.36	(7.48)	
Belgium (French)	36.06	(3.52)	485.24	(5.21)	4.13	(6.75)	
Germany	28.20	(3.59)	526.26	(6.17)	18.65	(7.53)	*
Finland	27.30	(2.62)	539.34	(5.20)	4.86	(5.18)	
France	27.13	(3.58)	500.82	(6.04)	7.02	(6.92)	
Ireland	24.23	(2.84)	544.82	(7.60)	11.24	(8.09)	
Belgium (Flemish)	22.60	(3.36)	488.35	(7.97)	4.17	(8.13)	
Netherlands	19.91	(4.09)	509.56	(8.12)	11.67	(9.20)	

Note: SE, standard error
*Significant difference between the mean scale scores at the time of teaching

Table 4.16 Percentage of students being taught two life science topics by their teachers

Education system		Relationships in simple food chains %	Common sources of energy %	Education system		Relationships in simple food chains %	Common sources of energy %
Albania	Before	23.5 (4.7)	24.7 (4.5)	Kosovo	Before	74.5 (2.9)	32.4 (3.2)
	This year	48.4 (5.8)	45.1 (5.5)		This year	23.9 (2.9)	65.0 (3.3)
	Not yet	28.1 (4.8)	30.2 (5.5)		Not yet	1.7 (0.8)	2.6 (0.8)
Belgium (Flemish)	Before	49.3 (5.5)	12.0 (3.2)	Latvia	Before	27.2 (2.4)	6.8 (1.4)
	This year	29.2 (4.7)	55.5 (4.7)		This year	64.1 (2.4)	81.0 (2.4)
	Not yet	21.5 (4.0)	32.4 (4.3)		Not yet	8.7 (1.5)	12.2 (2.0)
Belgium (French)	Before	69.1 (3.1)	17.2 (2.7)	Lithuania	Before	12.7 (3.9)	12.3 (3.3)
	This year	30.4 (3.1)	81.4 (2.8)		This year	52.6 (5.8)	66.0 (5.9)
	Not yet	0.5 (0.5)	1.4 (0.7)		Not yet	34.8 (5.6)	21.6 (4.8)
Bosnia and Herzegovina	Before	68.3 (3.1)	11.5 (2.1)	Montenegro	Before	6.6 (1.7)	3.0 (1.2)
	This year	26.4 (3.3)	66.2 (3.2)		This year	85.5 (2.6)	21.2 (2.9)
	Not yet	5.4 (1.8)	22.3 (2.7)		Not yet	7.9 (2.2)	75.8 (2.9)
Bulgaria	Before	29.5 (2.7)	9.2 (1.9)	Netherlands	Before	82.9 (2.4)	29.4 (2.8)
	This year	56.8 (3.1)	69.4 (3.2)		This year	13.6 (2.3)	57.4 (3.2)
	Not yet	13.7 (2.3)	21.4 (2.8)		Not yet	3.5 (1.0)	13.2 (2.3)
Cyprus	Before	33.1 (3.8)	14.2 (2.6)	N. Macedonia, Rep. of	Before	26.2 (4.2)	42.1 (5.3)
	This year	47.2 (4.8)	55.3 (4.2)		This year	62.7 (5.0)	55.0 (5.6)
	Not yet	19.7 (3.7)	30.4 (4.2)		Not yet	11.1 (4.1)	2.9 (1.9)
Czech Republic	Before	44.1 (2.9)	15.7 (2.4)	Poland	Before	86.1 (2.4)	43.0 (4.0)
	This year	46.3 (3.0)	52.0 (3.5)		This year	13.5 (2.4)	55.4 (3.9)
	Not yet	9.5 (1.6)	32.3 (3.3)		Not yet	0.4 (0.4)	1.6 (0.7)

(continued)

Table 4.16 (continued)

Education system		Relationships in simple food chains %	Common sources of energy %	Education system		Relationships in simple food chains %	Common sources of energy %
Denmark	Before	47.0 (4.0)	8.5 (1.9)	Portugal	Before	18.8 (2.4)	36.6 (3.0)
	This year	30.2 (3.7)	62.3 (4.0)		This year	57.1 (3.4)	36.1 (3.0)
	Not yet	22.8 (3.5)	29.2 (3.6)		Not yet	24.1 (3.3)	27.3 (3.2)
Finland	Before	47.7 (3.7)	29.0 (3.3)	Romania	Before	27.9 (3.2)	7.9 (2.1)
	This year	24.9 (3.2)	45.6 (3,7)		This year	24.5 (2.8)	78.3 (2.9)
	Not yet	27.4 (3.5)	25.3 (3,0)		Not yet	47.6 (3.4)	13.8 (2.3)
France	Before	68.5 (3.6)	20.1 (3,1)	Serbia	Before	33.1 (3.0)	24.5 (2.4)
	This year	17.8 (2.7)	64.3 (4.0)		This year	42.0 (2.7)	52.6 (3.2)
	Not yet	13.7 (2.8)	15.5 (3.2)		Not yet	24.9 (2.3)	22.9 (2.4)
Germany	Before	36.0 (4.0)	22.9 (3.2)	Slovak Republic	Before	15.6 (3.3)	3.9 (1.6)
	This year	36.7 (3.8)	69.5 (3.4)		This year	68.8 (3.9)	49.9 (4.1)
	Not yet	27.3 (3.0)	7.6 (1.5)		Not yet	15.6 (3.1)	46.2 (4.0)
Hungary	Before	32.0 (3.3)	13.9 (2.3)	Slovenia	Before	25.4 (3.8)	21.4 (3.0)
	This year	39.5 (3.0)	39.6 (3.6)		This year	55.9 (4.5)	72.5 (3.2)
	Not yet	28.4 (3.2)	46.5 (3.7)		Not yet	18.7 (3.1)	6.2 (1.7)
Ireland	Before	33.9 (4.5)	16.9 (3.1)	Spain	Before	21.1 (3.3)	11.2 (2.2)
	This year	43.8 (4.2)	79.9 (3.5)		This year	41.0 (3.8)	49.5 (4.4)
	Not yet	22.3 (3.4)	3.2 (2.0)		Not yet	37.9 (3.2)	39.4 (3.8)
Italy	Before	44.4 (3.9)	35.0 (4.1)	Sweden	Before	23.7 (3.0)	7.9 (1.8)
	This year	27.2 (3.2)	23.9 (3.3)		This year	43.0 (3.3)	24.0 (2.8)
	Not yet	28.4 (3.7)	41.0 (3.9)		Not yet	33.3 (3.2)	68.1 (3.2)

Note: Standard errors appear in parentheses

Table 4.16 (continued)

Education system		Relation-ships in simple food chains		Common sources of energy			Relation-ships in simple food chains		Common sources of energy	
		%	SE	%	SE		%	SE	%	SE
Albania	Before	23.5	(4.7)	24.7	(4.5)	Before	74.5	(2.9)	32.4	(3.2)
	This year	48.4	(5.8)	45.1	(5.5)	This year	23.9	(2.9)	65.0	(3.3)
	Not yet	28.1	(4.8)	30.2	(5.5)	Not yet	1.7	(0.8)	2.6	(0.8)
Belgium (Flemish)	Before	49.3	(5.5)	12.0	(3.2)	Before	27.2	(2.4)	6.8	(1.4)
	This year	29.2	(4.7)	55.5	(4.7)	This year	64.1	(2.4)	81.0	(2.4)
	Not yet	21.5	(4.0)	32.4	(4.3)	Not yet	8.7	(1.5)	12.2	(2.0)
Belgium (French)	Before	69.1	(3.1)	17.2	(2.7)	Before	12.7	(3.9)	12.3	(3.3)
	This year	30.4	(3.1)	81.4	(2.8)	This year	52.6	(5.8)	66.0	(5.9)
	Not yet	0.5	(0.5)	1.4	(0.7)	Not yet	34.8	(5.6)	21.6	(4.8)
Bosnia and Herzegovina	Before	68.3	(3.1)	11.5	(2.1)	Before	6.6	(1.7)	3.0	(1.2)
	This year	26.4	(3.3)	66.2	(3.2)	This year	85.5	(2.6)	21.2	(2.9)
	Not yet	5.4	(1.8)	22.3	(2.7)	Not yet	7.9	(2.2)	75.8	(2.9)
Bulgaria	Before	29.5	(2.7)	9.2	(1.9)	Before	82.9	(2.4)	29.4	(2.8)
	This year	56.8	(3.1)	69.4	(3.2)	This year	13.6	(2.3)	57.4	(3.2)
	Not yet	13.7	(2.3)	21.4	(2.8)	Not yet	3.5	(1.0)	13.2	(2.3)
Cyprus	Before	33.1	(3.8)	14.2	(2.6)	Before	26.2	(4.2)	42.1	(5.3)
	This year	47.2	(4.8)	55.3	(4.2)	This year	62.7	(5.0)	55.0	(5.6)
	Not yet	19.7	(3.7)	30.4	(4.2)	Not yet	11.1	(4.1)	2.9	(1.9)
Czech Republic	Before	44.1	(2.9)	15.7	(2.4)	Before	86.1	(2.4)	43.0	(4.0)
	This year	46.3	(3.0)	52.0	(3.5)	This year	13.5	(2.4)	55.4	(3.9)
	Not yet	9.5	(1.6)	32.3	(3.3)	Not yet	0.4	(0.4)	1.6	(0.7)
Denmark	Before	47.0	(4.0)	8.5	(1.9)	Before	18.8	(2.4)	36.6	(3.0)
	This year	30.2	(3.7)	62.3	(4.0)	This year	57.1	(3.4)	36.1	(3.0)
	Not yet	22.8	(3.5)	29.2	(3.6)	Not yet	24.1	(3.3)	27.3	(3.2)
Finland	Before	47.7	(3.7)	29.0	(3.3)	Before	27.9	(3.2)	7.9	(2.1)
	This year	24.9	(3.2)	45.6	(3,7)	This year	24.5	(2.8)	78.3	(2.9)
	Not yet	27.4	(3.5)	25.3	(3,0)	Not yet	47.6	(3.4)	13.8	(2.3)

(continued)

Table 4.16 (continued)

Education system		Relation-ships in simple food chains		Common sources of energy			Relation-ships in simple food chains		Common sources of energy	
		%	SE	%	SE		%	SE	%	SE
France	Before	68.5	(3.6)	20.1	(3,1)	Before	33.1	(3.0)	24.5	(2.4)
	This year	17.8	(2.7)	64.3	(4.0)	This year	42.0	(2.7)	52.6	(3.2)
	Not yet	13.7	(2.8)	15.5	(3.2)	Not yet	24.9	(2.3)	22.9	(2.4)
Germany	Before	36.0	(4.0)	22.9	(3.2)	Before	15.6	(3.3)	3.9	(1.6)
	This year	36.7	(3.8)	69.5	(3.4)	This year	68.8	(3.9)	49.9	(4.1)
	Not yet	27.3	(3.0)	7.6	(1.5)	Not yet	15.6	(3.1)	46.2	(4.0)
Hungary	Before	32.0	(3.3)	13.9	(2.3)	Before	25.4	(3.8)	21.4	(3.0)
	This year	39.5	(3.0)	39.6	(3.6)	This year	55.9	(4.5)	72.5	(3.2)
	Not yet	28.4	(3.2)	46.5	(3.7)	Not yet	18.7	(3.1)	6.2	(1.7)
Ireland	Before	33.9	(4.5)	16.9	(3.1)	Before	21.1	(3.3)	11.2	(2.2)
	This year	43.8	(4.2)	79.9	(3.5)	This year	41.0	(3.8)	49.5	(4.4)
	Not yet	22.3	(3.4)	3.2	(2.0)	Not yet	37.9	(3.2)	39.4	(3.8)
Italy	Before	44.4	(3.9)	35.0	(4.1)	Before	23.7	(3.0)	7.9	(1.8)
	This year	27.2	(3.2)	23.9	(3.3)	This year	43.0	(3.3)	24.0	(2.8)
	Not yet	28.4	(3.7)	41.0	(3.9)	Not yet	33.3	(3.2)	68.1	(3.2)

Table 4.17 Teaching of Earth science topics and students' environmental mean scale scores

Education system		Earth science: Physical characteristics of Earth					Earth science: Earth's rising average temperatures and results of this change				
		%	SE	Mean	SE	Sig.	%	SE	Mean	SE	Sig.
Albania	Before	33.7	(5.6)	489.0	(10.5)		25.1	(5.5)	472.4	(18.3)	
	This year	21.5	(4.3)	490.5	(10.2)		29.2	(5.4)	487.9	(9.9)	
	Not yet	44.9	(5.8)	465.9	(11.8)		45.7	(4.9)	476.9	(8.1)	
Belgium (Flemish)	Before	8.8	(2.5)	495.1	(6.5)		4.5	(1.5)	499.2	(12.1)	
	This year	23.0	(3.6)	502.8	(5.7)		50.3	(4.1)	494.9	(5.2)	
	Not yet	68.2	(3.8)	487.7	(4.0)	*	45.2	(4.2)	487.6	(5.2)	
Belgium (French)	Before	8.5	(2.1)	475.4	(13.7)		7.8	(1.6)	473.9	(11.1)	
	This year	18.5	(3.0)	472.9	(6.3)		23.9	(3.0)	477.5	(6.6)	
	Not yet	73.0	(3.2)	485.7	(3.5)		68.3	(3.1)	485.2	(3.9)	
Bosnia and Herzegovina	Before	13.5	(3.5)	456.0	(9.8)		0.8	(0.7)	487.0	(49.7)	
	This year	41.4	(5.8)	456.7	(6.9)		14.7	(3.4)	453.4	(9.4)	
	Not yet	45.1	(5.7)	439.1	(8.6)		84.5	(3.4)	447.0	(5.4)	
Bulgaria	Before	34.2	(3.6)	565.3	(8.8)		15.4	(2.9)	552.6	(17.1)	
	This year	55.9	(3.5)	522.8	(9.7)		36.9	(4.1)	535.1	(11.2)	
	Not yet	9.9	(2.0)	491.5	(25.8)	*	47.7	(4.3)	529.8	(9.7)	
Cyprus	Before	36.1	(3.2)	483.5	(4.4)		12.0	(2.6)	487.2	(10.7)	
	This year	13.9	(2.4)	493.5	(12.6)		24.1	(2.9)	472.3	(6.4)	
	Not yet	50.0	(3.6)	473.1	(4.8)		63.8	(3.2)	481.3	(3.6)	
Czech Republic	Before	16.7	(2.4)	516.0	(5.9)		4.4	(1.0)	512.4	(8.8)	
	This year	63.4	(2.9)	521.6	(3.6)		29.1	(3.0)	519.8	(5.8)	
	Not yet	19.9	(2.6)	518.1	(4.7)		66.5	(3.0)	519.3	(3.1)	
Denmark	Before	20.9	(3.6)	541.1	(7.0)		12.9	(2.8)	533.4	(7.4)	
	This year	44.7	(4.6)	536.5	(4.6)		54.3	(4.1)	538.2	(4.9)	
	Not yet	34.4	(4.2)	533.2	(5.6)		32.8	(4.0)	534.5	(5.6)	

(continued)

Table 4.17 (continued)

Education system		Earth science: Physical characteristics of Earth					Earth science: Earth's rising average temperatures and results of this change				
		%	SE	Mean	SE	Sig.	%	SE	Mean	SE	Sig.
Finland	Before	29.7	(2.8)	550.0	(4.5)		9.9	(1.8)	540.8	(8.0)	
	This year	47.9	(3.5)	541.9	(4.6)		37.1	(2.6)	548.1	(4.6)	
	Not yet	22.4	(2.8)	534.1	(5.7)		53.1	(2.5)	539.4	(3.7)	
France	Before	27.1	(3.2)	496.8	(6.0)		5.8	(2.7)	507.0	(19.5)	
	This year	35.5	(3.8)	496.5	(5.7)		32.2	(3.3)	493.2	(4.2)	
	Not yet	37.4	(4.3)	494.7	(6.9)		62.0	(3.9)	496.3	(5.4)	
Germany	Before	19.7	(3.3)	515.1	(6.9)		23.9	(3.2)	511.5	(5.5)	
	This year	37.4	(3.5)	521.2	(5.1)		37.0	(3.4)	518.7	(5.8)	
	Not yet	42.8	(3.5)	507.7	(5.6)		39.1	(3.5)	508.6	(5.3)	
Hungary	Before	21.9	(3.5)	537.1	(10.8)		7.8	(1.8)	564.0	(8.3)	
	This year	60.1	(4.1)	534.2	(5.6)		48.4	(4.1)	535.8	(5.3)	
	Not yet	18.0	(3.3)	523.9	(9.3)		43.8	(4.0)	524.4	(7.4)	*
Ireland	Before	26.5	(3.5)	537.5	(5.8)		13.2	(2.6)	523.4	(11.0)	
	This year	48.9	(4.4)	536.4	(5.9)		68.8	(3.3)	540.5	(4.6)	
	Not yet	24.7	(3.8)	533.1	(6.6)		18.0	(3.0)	528.3	(6.7)	
Italy	Before	28.3	(3.1)	518.6	(4.5)		12.9	(2.3)	509.6	(8.2)	
	This year	60.4	(3.6)	515.9	(3.9)		57.2	(3.9)	521.7	(3.5)	
	Not yet	11.3	(2.1)	523.1	(7.6)		29.9	(3.7)	512.9	(5.8)	
Kosovo	Before	22.5	(3.6)	404.8	(7.3)		12.5	(2.6)	392.2	(5.7)	
	This year	51.2	(4.3)	394.4	(6.0)		48.0	(4.4)	392.3	(6.5)	
	Not yet	26.3	(3.8)	415.9	(6.6)	*	39.6	(4.3)	417.4	(5.6)	*
Latvia	Before	52.3	(4.3)	505.5	(4.5)		21.7	(3.3)	510.9	(5.8)	
	This year	11.7	(2.6)	505.8	(9.4)		20.9	(3.5)	508.3	(7.8)	
	Not yet	36.0	(3.9)	500.5	(5.6)		57.4	(4.1)	500.2	(4.6)	

(continued)

Table 4.17 (continued)

Education system		Earth science: Physical characteristics of Earth					Earth science: Earth's rising average temperatures and results of this change				
		%	SE	Mean	SE	Sig.	%	SE	Mean	SE	Sig.
Lithuania	Before	52.2	(3.5)	522.2	(4.6)		34.2	(3.2)	518.7	(6.3)	
	This year	45.2	(3.6)	518.0	(4.8)		50.8	(3.7)	521.5	(3.9)	
	Not yet	2.6	(1.0)	534.2	(16.0)		14.9	(2.6)	521.8	(7.6)	
Montenegro	Before	18.0	(2.4)	463.6	(6.3)		4.1	(0.8)	452.3	(11.1)	
	This year	55.7	(2.9)	466.9	(3.5)		21.5	(2.4)	479.7	(5.6)	
	Not yet	26.3	(2.9)	464.3	(5.1)		74.5	(2.5)	462.4	(2.6)	*
Netherlands	Before	19.0	(4.5)	528.0	(6.7)		6.0	(2.3)	497.9	(15.2)	
	This year	41.1	(5.3)	514.8	(4.2)		60.5	(4.7)	522.0	(3.7)	
	Not yet	39.9	(4.7)	519.1	(5.2)		33.5	(4.3)	517.3	(5.6)	
N. Macedonia, Rep. of	Before	22.2	(3.2)	426.9	(10.3)		17.0	(2.9)	427.4	(11.1)	
	This year	64.3	(3.6)	439.6	(6.7)		53.3	(4.1)	435.7	(7.8)	
	Not yet	13.5	(2.4)	442.7	(16.5)		29.7	(3.5)	446.0	(10.8)	
Poland	Before	4.8	(1.6)	552.2	(10.3)		0.8	(0.8)	539.8	(0.0)	
	This year	79.9	(3.5)	557.6	(2.8)		12.5	(2.3)	555.1	(5.4)	
	Not yet	15.3	(3.1)	555.6	(7.1)		86.7	(2.5)	557.4	(2.6)	
Portugal	Before	35.0	(3.3)	512.8	(5.1)		36.7	(2.7)	514.0	(5.3)	
	This year	63.1	(3.3)	521.9	(4.4)		52.8	(2.9)	520.3	(4.3)	
	Not yet	1.9	(1.1)	491.8	(10.4)		10.5	(2.0)	522.0	(10.9)	
Romania	Before	40.0	(5.2)	530.4	(7.3)		25.7	(4.5)	524.0	(7.4)	
	This year	52.3	(4.8)	521.5	(7.8)		68.1	(4.7)	527.3	(7.2)	
	Not yet	7.7	(2.7)	527.6	(23.6)		6.2	(2.4)	512.9	(20.8)	

(continued)

Table 4.17 (continued)

Education system		Earth science: Physical characteristics of Earth					Earth science: Earth's rising average temperatures and results of this change				
		%	SE	Mean	SE	Sig.	%	SE	Mean	SE	Sig.
Serbia	Before	57.2	(4.0)	519.1	(3.9)		22.5	(3.0)	517.6	(4.9)	
	This year	22.2	(3.4)	514.9	(7.2)		25.8	(3.5)	515.9	(6.5)	
	Not yet	20.6	(3.1)	514.0	(6.7)		51.6	(4.1)	518.8	(4.7)	
Slovak Republic	Before	31.3	(3.0)	522.4	(6.6)		11.1	(2.0)	531.8	(9.5)	
	This year	25.3	(2.9)	508.5	(9.6)		30.5	(2.6)	518.8	(7.0)	
	Not yet	43.3	(3.2)	499.6	(6.9)		58.3	(3.1)	499.6	(4.9)	*
Slovenia	Before	23.4	(2.8)	528.6	(5.0)		3.0	(1.1)	519.0	(10.8)	
	This year	47.8	(3.7)	517.9	(3.5)		28.6	(3.3)	524.5	(4.1)	
	Not yet	28.8	(3.2)	517.9	(4.2)	*	68.4	(3.5)	518.9	(3.3)	
Spain	Before	39.3	(3.2)	511.2	(3.5)		19.7	(2.3)	508.1	(4.3)	
	This year	53.4	(3.1)	508.0	(3.7)		58.5	(2.9)	512.1	(3.6)	
	Not yet	7.2	(1.4)	519.7	(7.2)		21.7	(2.3)	506.8	(4.1)	
Sweden	Before	8.9	(1.9)	535.4	(7.2)		7.4	(2.3)	527.4	(9.5)	
	This year	53.5	(4.5)	524.8	(4.6)		52.9	(4.1)	534.2	(3.9)	
	Not yet	37.6	(4.4)	535.7	(5.2)		39.7	(4.3)	524.2	(5.2)	

Note: Standard errors appear in parentheses
*Significant difference between the mean scale scores at the time of teaching

Table 4.18 How often teachers do environmental activities in classes and student environmental knowledge

Education system	Develop students' positive attitudes toward the natural environment	Encourage students to use less resources	Discuss how student actions in and outside of school help the natural environment	Discuss environmental issues
Poland	◉	○	○	○
Finland	◉	○	○	○
Ireland	○	◉	○	○
Denmark	○	○	○	○
Hungary	◉	◉	◉	○
Bulgaria	◉	◉	◉	◉
Sweden	○	○	○	○
Romania	◉	◉	◉	◉
Lithuania	◉	◉	○	○
Czech Republic	◉	○	○	○
Slovenia	◉	◉	◉	○
Netherlands	○	○	○	○
Portugal	◉	◉	◉	○
Italy	◉	◉	◉	◉
Serbia	◉	◉	◉	○
Germany	◉	○	○	○
Spain	◉	◉	◉	○
Slovak Republic	◉	◉	○	○
Latvia	◉	○	○	○
France	○	○	○	○
Belgium (Flemish)	○	○	○	○
Belgium (French)	○	○	○	○
Cyprus	◉	○	○	○
Albania	○	○	○	○
Montenegro	◉	◉	◉	○
Bosnia and Herzegovina	◉	◉	◉	○
NM Macedonia, Rep. of	◉	◉	◉	◉
Kosovo	◉	○	◉	◉
Table average	◉	○	○	○

Notes: Education systems are ordered according to the national mean student achievement in environmental knowledge; see Table 4.2; ◉, at least once a week; ○, once or twice a month (Other possible options were a few times a year, or never/almost never.)

References

Baierl, T.-M., Kaiser, F. G., & Bogner, F. X. (2024). The role of attitude toward nature in learning about environmental issues. *Frontiers in Psychology, 15*, 1471026. https://doi.org/10.3389/fpsyg.2024.1471026

Bianchi, G., Pisiotis, U., & Cabrera Giraldez, C. (2022). *GreenComp: The European sustainability competence framework*. European Commission: Joint Research Centre. Publications Office of the European Union. https://doi.org/10.2760/13286

Cebrián, G., Corres Gallardo, A. d. C., Renta Davids, A. I., & Junyent, M. (2025). Assessment of sustainability competencies in primary education: A systematic review. *The Journal of Environmental Education, 56*(3), 208–228. https://doi.org/10.1080/00958964.2025.2471967.

Corres, A., Rieckmann, M., Espasa, A., & Ruiz-Mallén, I. (2020). Educator competences in sustainability education: A systematic review of frameworks. *Sustainability, 12*(23), 9858. https://doi.org/10.3390/su12239858

Fishbein, B., Taneva, M., & Kowolik, K. (2025). *TIMSS 2023 international database*. TIMSS & PIRLS International Study Center, Boston College. https://timss2023.org/data/

IBM. (2021). *IBM SPSS software*. https://www.ibm.com/products/spss-statistics.

IEA. (2021). *Data and tools. Tools*. https://www.iea.nl/data-tools/tools.

Kácovský, P., Jedličková, T., Kuba, R., Snětinová, M., Surynková, P., Vrhel, M., & Urválková, E. S. (2021). Lower secondary intended curricula of science subjects and mathematics: a comparison of the Czech Republic, Estonia, Poland and Slovenia. *Journal of Curriculum Studies, 54*(3), 384–405. https://doi.org/10.1080/00220272.2021.1978557

Kioupi, V., & Voulvoulis, N. (2019). Education for sustainable development: A systemic framework for connecting the SDGs to educational outcomes. *Sustainability, 11*(21), 6104. https://doi.org/10.3390/su11216104

Kovách, I., Megyesi, B. G., Barthes, A., Oral, H. V., & Smederevac-Lalic, M. (2021). Knowledge use in education for environmental citizenship—Results of four case studies in Europe (France, Hungary, Serbia, Turkey). *Sustainability, 13*(19), 11118. https://doi.org/10.3390/su131911118

Krayneva, R., Rudenko, A., & Motylev, R. (2021). Role of education in implementing the sustainable development strategy. *E3S Web of Conferences, 250*, 07008. https://doi.org/10.1051/e3sconf/202125007008

Liefländer, A. K., & Bogner, F. X. (2014). The effects of children's age and sex on acquiring pro-environmental attitudes through environmental education. *The Journal of Environmental Education, 45*(2), 105–117. https://doi.org/10.1080/00958964.2013.875511

Momete, D. C., & Momete, M. M. (2021). Map and track the performance in education for sustainable development across the European Union. *Sustainability, 13*(23), 13185. https://doi.org/10.3390/su132313185

Mullis, I. V. S., Martin, M. O., & von Davier, M. (Eds.). (2021). *TIMSS 2023 assessment frameworks*. TIMSS & PIRLS International Study Center, Boston College. https://timssandpirls.bc.edu/timss2023/frameworks/index.html

Reynolds, K. A., Aldrich, C. E. A., Bookbinder, A., Gallo, A., von Davier, M., & Kennedy, A. (Eds.). (2024). *TIMSS 2023 encyclopedia: Education policy and curriculum in mathematics and science*. TIMSS & PIRLS International Study Center, Boston College. https://doi.org/10.6017/lse.tpisc.timss.rs5882

Reynolds, K. A., & Komakhidze, M. (2022). *TIMSS 2023 environmental attitudes and behaviors framework*. TIMSS & PIRLS International Study Center, Boston College. https://timssandpirls.bc.edu/timss2023/frameworks/pdf/T23-Environmental-Attitudes-and-Behaviors-Framework.pdf

UNESCO. (2020). *Education for sustainable development: A roadmap*. UNESCO. https://doi.org/10.54675/YFRE1448

United Nations. (2015). *Transforming our world: The 2030 agenda for sustainable development* (A/RES/70/1). https://sdgs.un.org/2030agenda.

von Davier, M., Fishbein, B., & Kennedy, A. (Eds.). (2024). *TIMSS 2023 technical report (methods and procedures)*. TIMSS & PIRLS International Study Center, Boston College. https://timss2023.org/methods

von Davier, M., Kennedy, A., Reynolds, K., Fishbein, B., Khorramdel, L., Aldrich, C., Bookbinder, A., Bezirhan, U., & Yin, L. (2024). *TIMSS 2023 international results in mathematics and science*. TIMSS & PIRLS International Study Center, Boston College. https://doi.org/10.6017/lse.tpisc.timss.rs6460

Vukić, T., Jovanović, M., & Todorović, D. (2021). Goals and objectives of education for sustainable development as modern curriculum innovation in Serbia, Montenegro and Croatia. *Facta Universitatis, Series: Philosophy, Sociology, Psychology and History, 20*(1), 55–72. https://doi.org/10.22190/FUPSPH2101055V

Barbara Japelj Pavešič is a researcher at the Educational Research Institute Ljubljana, Slovenia. She is involved in studies of TIMSS from 1995 and from 2003, TIMSS national research coordinator. She was a member of the Questionnaire Item Review Committee (QIRC) for TIMSS 2023 and 2027. Her main interest is student knowledge, mathematics and science, of students from kindergarten to pre-university education and especially research of motivation and system factors that can be improved to allow better learning and higher knowledge of students. She is recently also in position of national project manager for the study of learning and teaching, TALIS and involved in other national projects and studies, for which she is responsible to write national reports and conduct secondary analyses of the nationally important issues and findings.

Jure Novak is a research assistant at the Centre for Applied Epistemology at the Educational Research Institute, based in Ljubljana, Slovenia. He primarily works as a data manager on international large-scale assessments (ILSAs), such as IEA's PIRLS and TIMSS, as well as OECD's TALIS. His research interests include the sociology of education, education for sustainable development, and quantitative methodology.

Chapter 5
Teaching Environmental Sustainability in the Science Classroom: Lessons from and for the TIMSS Environmental Framework

Jelle Boeve-de Pauw, Michiel van Harskamp, Wanda Sass, Elisa Caponera, Riccardo Pietracci, and Laura Palmerio

5.1 Introduction

The previous chapters of this volume have illustrated how environmental knowledge is operationalized in the Trends in International Mathematics and Science Study (TIMSS) 2023 and how education systems differ in terms of their students' environmental knowledge and achievement scores. Environmental education, however, concerns more than environmental knowledge, and several authors have defined differences within environmental knowledge (as learning goals for environmental education) that are of particular interest. These differences highlight, for instance, that beyond theoretical knowledge of environmental issues (for example, knowing what causes ocean acidification), more applied forms of knowledge, pertaining to the impact of choices people make on the environment (for example, knowing how personal or collective mobility contributes more or less to environmental pollution), are also important to consider as outcomes of sustainability education (Roczen et al., 2014). Besides knowledge, sustainability education typically focuses on attitudinal and behavioral student outcomes, and data from international large-scale assessments have been used before to study how these differ between and within education systems (Boeve-de Pauw & Van Petegem, 2010; Coertjens et al., 2010; Gong & Zheng, 2021; Isac et al., 2025; List et al., 2020). Using TIMSS 2023 data, this chapter focuses on activities that primary school teachers report implementing in the (science) classroom, specifically related to teaching

J. B.-d. Pauw (✉) · M. van Harskamp · W. Sass
Freudenthal Institute, Utrecht University, Utrecht, The Netherlands
e-mail: j.n.a.boeve-depauw@uu.nl; m.vanharskamp@uu.nl; w.p.y.sass@uu.nl

E. Caponera · R. Pietracci · L. Palmerio
Istituto Nazionale Per la Valutazione del Sistema Educativo di Istruzione e di Formazione (INVALSI), Rome, Italy
e-mail: elisa.caponera@invalsi.it; riccardo.pietracci@invalsi.it; laura.palmerio@invalsi.it

D. Hastedt et al. (eds.), *Educating Pupils for Environmental Sustainability in Europe*, IEA Research for Education 18,
https://doi.org/10.1007/978-3-032-14725-7_5

environmental sustainability, to explore how these teaching activities could contribute to students' attitudes and behaviors.

Previous studies within single education systems show how teaching approaches that teachers deploy in the (science) classroom can have an impact on their students' environmental attitudes and behaviors (Boeve-de Pauw & Van Petegem, 2018; Yeung, 2004). Often, studies that establish such connections between what teachers do in the classroom (teaching activities and/or approaches) on the one hand and student outcomes in terms of environmental attitudes and behaviors on the other, are performed in specific educational contexts and using specific samples that do not allow findings to be generalized to the level of education systems. The TIMSS data and the associated environmental framework present a unique opportunity to go beyond such specific educational contexts and explore issues at the education system level. The potential of these data to contribute to the wider research fields of environmental and sustainability education are explored.

The current chapter, therefore, focuses on the data that were collected using the teacher surveys within TIMSS, concerning the activities teachers deploy to bring environmental sustainability into their (science) education. After describing this information for the education systems that are included in the current volume, data about teachers' reported activities is connected to students' environmental attitudes and behaviors as captured in the TIMSS 2023 student questionnaires. The aim is to explore the potential of these data to contribute to the wider research fields of environmental and sustainability education. To achieve this, relevant sections of the TIMSS questionnaires are first presented, followed by a description of how contemporary research, specifically in the field of sustainability education, understands the goals of and approaches to such activities.

5.1.1 The TIMSS 2023 Environmental Attitudes and Behaviors Framework

In TIMSS 2023, attitudes and behaviors are assessed using the student questionnaire, with all items based on the TIMSS 2023 Environmental Attitudes and Behaviors Framework (Reynolds & Komakhidze, 2022). To measure students' environmental attitudes, TIMSS 2023 draws from the two-dimensional Model of Ecological Values, which is based on the Theory of Ecological Attitudes (Bogner & Wiseman, 1999, 2006; Wiseman & Bogner, 2003). TIMSS 2023 items that build on this model include, for example, "I care about the protection of plants and animals" and "Addressing climate change should be a high priority," reflecting an individual's endorsement of conservation and protection of nature. Both these example items can be found in question G11 in the student questionnaire. Students are asked to indicate how much they agree with such statements. Those students whose attitudes are aligned with preservation are likely to enjoy spending time in nature and care about the protection of natural areas. Measuring environmentally responsible behavior is identified as a complex task by the TIMSS 2023 framework, as children in primary education tend to have limited agency in making decisions and their ability to exercise control likely varies substantially across education systems, cultures, and households. Due to these complications, TIMSS 2023 student questionnaire items measuring students' environmentally responsible behaviors reflect relatively simple practices that are accessible to students across a variety of contexts. Examples include: "I try to reuse things (e.g., bags, bottles…)" or "I tell my friends when they are doing things that harm the environment," which are both part of question G12 of the student questionnaire. Students indicate how often they engage in these behaviors.

The TIMSS 2023 Environmental Attitudes and Behaviors Framework also highlights the importance of the classroom context (Reynolds & Komakhidze, 2022). It notes that students who experience more active and interactive teaching methods—such as classroom discussions, research projects, hands-on activities, excursions in nature, outdoor environmental programs—demonstrate higher awareness about environmental issues, more positive environmental attitudes, and more frequent environmentally responsible behavior, based on insights from research literature (see, for example, Boeve-de Pauw et al., 2015; Krnel & Naglic, 2009).

These classroom activities are the main focus of the current chapter and are examined through three questions in the TIMSS 2023 teacher questionnaire (Table 5.1). One question (S6) deals with the extent to which teachers perceive that education for environmental sustainability should be a priority in schools. The two other questions focus on the concrete activities that teachers report doing in their

Table 5.1 Questions S6, S4 and S5 from the grade four TIMSS teacher questionnaire

	Stem and items	Response options
S6	How much do you agree or disagree that education about environmental sustainability should be a priority for schools?	– Agree a lot – Agree a little – Disagree a little – Disagree a lot
S4	How often do you do the following when teaching this class? a. Develop students' positive attitudes toward the natural environment b. Encourage students to use less resources (e.g., <water, energy>) c. Discuss how student actions in and outside of school can help the natural environment d. Discuss environmental issues (e.g., <climate change, endangered animals>)	– At least once a week – Once or twice a month – A few times a year – Never or almost never
S5	Do you do these things to teach students about environmental issues and sustainability? a. Take students to visit natural areas (e.g., <a pond or meadow>) b. Have students participate in environmentally responsible activities (e.g., <pick up trash>) c. Have students do research or projects on a particular environmental topic (e.g., <pollution, climate change>) d. Provide opportunities for students to participate in outdoor environmental education programs outside of school	– Yes – No

classrooms or school. Question S4 asks how often teachers perform activities that develop their students' positive attitudes towards the natural environment, or how often they discuss environmental issues in class. Question S5 asks whether teachers perform certain activities with their students—for instance, taking them to visit natural areas (for example, a pond or meadow), or conducting research projects on a particular environmental topic (for example, pollution or climate change)—using a yes or no response format.

The inclusion of these topics in both the student and teacher questionnaires enables exploration of how teaching environmental sustainability in primary school science classrooms is related to students' environmental attitudes and behavior.

It should be noted that, in contemporary research—particularly in the field of sustainability education—understandings of both learning goals and teaching approaches have evolved over the past decades. This evolution partially overlaps with the operationalization of the TIMSS 2023 Environmental Attitudes and Behaviors Framework (Reynolds & Komakhidze, 2022), but important differences exist that must be considered when exploring the potential that TIMSS presents for advancing the field of sustainability education.

5.1.2 From Behavior to Action, and from Teaching Activities to Action-Oriented Learning

Environmental education has traditionally been concerned with shaping pro-environmental behavior, while considering knowledge and attitudes as a means to achieve this goal (Macintyre et al., 2024). However, with the transition to sustainability education, this normative perspective of having learners make the "right" choice, has shifted the focus away from teaching the "right" behaviors. Emphasis has moved from promoting environmental behavior towards empowering learners in deliberate action taking, including fostering willingness to act (see, for example, United Nations Educational, Scientific and Cultural Organization [UNESCO], 2020) and adopting a learner-centered approach to teaching (European Commission, Directorate-General for Education, Youth, Sport and Culture, 2022). An example of this shift in focus from more closed behavioral outcomes to more open-ended action-taking outcomes can be seen in the European sustainability competence framework, GreenComp (Bianchi et al., 2022). As described elsewhere in this volume, GreenComp adopts *acting for sustainability* as one of its four competence areas, with three underlying competences integrated into it: *individual initiative, collective action* and *political agency* (Bianchi et al., 2022, p. 25). In this framework, a key goal for sustainability education is to encourage and invite learners to take action rather than perform pre-defined behaviors.

How behavior and action are understood fundamentally shapes their relation to education. Within sustainability education, an *action* is defined as a voluntary behavior—one that individuals or groups choose freely, rather than being forced upon them by others, including peers, parents, or educators. Furthermore, action differs from a mere activity in that it is directed towards a certain goal. This goal is typically aiming to contribute to finding solutions to so-called "wicked problems," which involves different or conflicting perspectives on how to solve a complex problem (Jensen, 2000; Jensen & Schnack, 2006; Mogensen et al., 2009). If education aims to empower students so that they are well-equipped for taking action (as defined here), a desired learning outcome is *action competence*. Sass et al. (2020) defined action competence as the ability to find or create relevant knowledge, the willingness (that is, passion and commitment) to contribute to solving a wicked problem, and the self-efficacy for doing so (that is, having confidence in one's own capacities and in the impact of taking a certain action). In line with its focus on action rather than behavior, the GreenComp framework emphasizes learners' willingness to act as an important empowering facet of sustainability education:

> Individual initiative relies on someone knowing what types of action are possible, having confidence in their own potential to influence change (internal locus of control), and being willing to act. … However, individual initiative does not only rely on opportunities for action and someone's self-awareness and self-efficacy. It also has a strong attitudinal aspect – the willingness to act (Bianchi et al., 2022, p. 27).

If education is to foster students' action competence, it should be action-oriented. An action-oriented educational approach firmly embeds education at school within the (local) society around it (Sinakou et al., 2019; Varela-Losada et al., 2016). Students work on real-world wicked problems they select together with the teacher. Consequently, they learn alongside the teacher by doing—for instance, by trying out a certain action and learning from mistakes (Sass et al., 2023). The process can be supported through joint critical reflection on factors that influenced the action, positively or negatively (Knippels & van Harskamp, 2018). Ultimately, this approach may foster a learning process in which students focus on opportunities rather than on unalterable detrimental conditions (Hasslöf & Malmberg, 2015; Sinakou et al., 2019).

In a science education context, action-oriented science education resonates with Sjöström and Eilks' (2018) *Vision 3* on scientific literacy. Whereas *Vision 1* emphasizes epistemology, dealing with the scientific understanding of concepts, and *Vision 2* relates to everyday-life contexts and the usefulness of science, Vision 3 concerns science education for transformation. Both Vision 2 and Vision 3 highlight the relevance of science and scientific information in societal issues, such as sustainability, but "if Vision [2] focuses on everyday-life relevance, Vision [3] focuses more on problematized relevance for critical citizenship and sustainability" (Sjöström & Eilks, 2018, p. 78). In this sense, action-oriented science education for sustainability, with its open-ended and participatory character, aligns closely with Vision 3 on scientific literacy.

Defining action in this way has consequences for the selected content and the action-oriented educational approach. It means, for example, that activities selected by the teacher alone—such as having students perform a scientific experiment—cannot be considered action-oriented unless students participate in selecting the experiment and in applying it to learn about an issue that matters to them. Additionally, there should be conflicting perspectives on how to solve the problem or perform the task. The operationalization of sustainability teaching in the TIMSS 2023 teacher questionnaires (as presented in the previous section of this chapter) touches on several aspects related to the distinction between teaching "activities" and "action-oriented" teaching and learning. After presenting the empirical parts of this chapter, the discussion will return to these lenses and reflect on how they could present further opportunities to leverage TIMSS for advancing understanding of teaching and learning for environmental sustainability in education systems across Europe and beyond.

5.1.3 Research Questions

The current chapter aims to use TIMSS 2023 data to explore how teachers report integrating sustainability education into their classrooms and how this relates to their students' environmental attitudes and behaviors. In comparing education systems, this chapter addresses two guiding questions:

1. What do teachers report about their science education practices regarding environmental sustainability?
2. What is the association between these teaching practices and students' environmental attitudes and behaviors?

Based on exploration of these issues through the TIMSS 2023 Environmental Attitudes and Behaviors Framework (Reynolds & Komakhidze, 2022), the chapter reflects on the opportunities for TIMSS to offer insight into action and action-orientation across education systems in future cycles. If seized, these opportunities may further underscore the relevance of TIMSS data for practitioners, researchers, and policymakers that are involved in monitoring and evaluating sustainability education.

While the analyses concern all selected education systems for this volume, results for Italy and the Netherlands are highlighted specifically to provide concrete examples. These two systems were selected because the recently published Eurydice report shows that they adopt substantially different approaches concerning how sustainability competences are embedded in the top-level curricula (European Commission/EACEA [Education, Audiovisual and Culture Executive Agency]/Eurydice, 2024). The report indicates that Italy incorporates sustainability competences mainly in a cross-curricular fashion (valuing sustainability), in citizenship education (for example, adaptability, individual and collective action), and in subjects such as languages or arts (for example, promoting nature, systems thinking). In contrast, the Netherlands includes only one sustainability competence (political agency) in the top-level curriculum and only in citizenship education. In this chapter, these selected case education systems and their differences will serve as illustrative examples to help make findings concrete, rather than to imply causal relationships between top-level curricula, teacher choices, and student performance.

5.2 Methodology

5.2.1 Sample

The analyses presented in this chapter were conducted on TIMSS 2023 data for grade four students. TIMSS employs a two-stage sampling design (for a detailed description, see von Davier et al., 2024). Of the 29 selected education systems, 4877 schools participated in the study (see Chap. 1 for the selection process of education systems). The overall student sample comprised 139,335 students, of whom 49.60 percent were female. The average age of the students was 10.35 years (standard deviation = 0.57). This sample is representative of the fourth-grade students across education systems.[1] The overall student sample was representative of more than 4,100,000 students in the target population across education systems. In addition, 7564 teachers who taught science in the classes of the participating schools were also involved. Most teachers (85.7%) were female.[2] In terms of age distribution, 82.2 percent of teachers were between 30 and 59 years of age. Of the teachers, 26.3 percent reported science as their major field of study. It should be noted that the sample design did not allow for estimating the characteristics of the teacher population since the teachers included in this study were not a representative sample of the national teaching population: TIMSS selects a representative sample of students in fourth grade in each participating education system, and all teachers who teach these selected students are involved in the survey.[3] As TIMSS employs a classroom-based sampling strategy, the resulting data may not be generalizable to the entire population of teachers within an education system.[4]

5.2.2 Variables

In order to answer the research questions, questions S6, S4, and S5 from the teacher questionnaire (see Table 5.1) were selected for analysis, as they align most strongly with the concept of action-orientation (Reynolds & Komakhidze, 2022). These

[1] In Norway, students were assessed in the fifth grade. Romania did not satisfy guidelines for sample participation rates.

[2] Based on the number of teachers who responded to the question regarding their gender.

[3] Generally, teachers in the fourth grade are not specialized science teachers but rather classroom teachers.

[4] For this reason, student weights will always be used in the following analyses.

questions collect data on the topics of teacher prioritization of sustainability in their education, frequency of performed sustainability education activities, and incorporation of particular sustainability education activities, respectively.

The student pro-environmental attitude scale was calculated by International Association for the Evaluation of Educational Achievement (IEA) researchers (Yin & Reynolds, 2024). The scale comprised five items assessing students' attitudes towards the environment: "I care about the protection of plants and animals," "I enjoy finding out what kinds of plants and animals live in my surrounding area," "I enjoy being in nature (e.g., <forests, parks, deserts>)," "It makes me sad when nature is destroyed," and "Addressing climate change should be a high priority." Students were required to indicate their level of agreement or disagreement on a 4-point Likert-type scale, ranging from *agree a lot* to *disagree a lot*. The internal consistency of the scale was deemed adequate, with a median Cronbach's alpha of 0.68 across education systems.

In a similar fashion, a new scale for student pro-environmental behaviors was constructed, based on students' answers to questions asking about the frequency of doing six things related to helping the natural environment (reusing things, saving resources, talking about how to help the environment, learning about environmental problems, participating in activities to help the environment, and telling friends when they are doing things that harm the environment). Students had to respond on a 4-point Likert-type scale ranging from *never* to *every day*. To estimate this scale, a procedure similar to that used in the IEA TIMSS assessments was used (von Davier et al., 2024). Specifically, a partial credit model analysis was conducted using Mplus (Muthén & Muthén, 2017), applying an item response theory model.[5] The scale scores derived from the partial credit model calibration were subsequently transformed into a standardized metric with a mean of 10 and a standard deviation of two. The scale's internal consistency, measured using Cronbach's alpha, was satisfactory (median across education systems = 0.79).

Initially, each education system contributes equally to the dataset by using student "senate weights," which total 500 across all student data within country. However, cases with missing responses to one or more items were excluded from the analysis without any adjustments to the corresponding weights. Consequently, education systems with a higher proportion of missing data may end up contributing less to the final analysis than those with more complete responses.

Similarly, a teacher scale based on teachers' answers to the four questions in S4 was estimated, using the student senate weights. The Cronbach's alpha for the teacher scale S4 was satisfactory (median α across education systems = 0.84), so this scale was deemed internally valid. The S5 items appeared heterogeneous and did not imply a latent construct. This was confirmed by the low Cronbach's alpha value ($\alpha < 0.60$).

[5] Cases with missing values in one or more items were excluded from the analysis. For scale calibration, each education system contributed data weighted by "senate weights," ensuring a total weight of 500 per education system's student dataset.

5.2.3 *Analytical Strategy*

For all analyses presented herein, the student weight for both student and teacher variables were utilized, in accordance with the methodology employed by IEA, as only the student sample can be considered representative. The education system-specific descriptive analyses, encompassing both teacher and student variables presented above, were conducted using the IEA IDB (International Database) Analyzer software (IEA, 2021), developed by the IEA Data Processing and Research Center for analyzing data from all IEA surveys, employing adapted macros provided by IEA TIMSS.

To investigate the relationship between the teacher variables and students' attitude and behavior scales, a multi-step approach was implemented. In the initial phase, for the two sample education systems, Italy and the Netherlands, as well as for all participating education systems, the mean scores for each item in S4 and S5 of the teacher questionnaire were calculated. Subsequently, the mean scores for the two teacher scales were computed: one pertaining to the topics taught, and the other to the set of environment-related activities. Lastly, considering all education systems, the relationship between the individual teacher items and the two teacher scales with students' attitude and behavior scales were examined. To quantify these associations, single regression analyses were conducted, using the IEA IDB Analyzer. Furthermore, for descriptive purposes, national tertiles were computed for the S4 scale and student attitudes and behaviors scores compared across the teacher tertile groups. Due to the large sample size TIMSS 2023 presents, it was decided to accept statistical significance at the level of $p > 0.01$ (Lantz, 2013).

5.3 Results

The first research question concerns teachers' perspectives on sustainability education in science education. To answer this question, the analysis examines the extent to which teachers prioritize sustainability education, as well as the educational activities they employ during sustainability education. In TIMSS 2023, these areas are covered by variables S6, S4, and S5. The figures in this section represent the proportion of students whose teachers gave the corresponding answers.

The data from S6 reflects the extent to which teachers think sustainability education should be prioritized in their school (Fig. 5.1). The data show that a vast majority of students in all education systems have teachers who see sustainability education as at least a partial priority of their teaching. Even in the education system with the biggest proportion of students with teachers who feel that sustainability education should not be a priority (the Netherlands), there is still a substantially larger proportion of students with teachers who think it should be a priority. Differences can, however, be found in the level of agreement within this subgroup of teachers who think sustainability education should be prioritized. The percentage

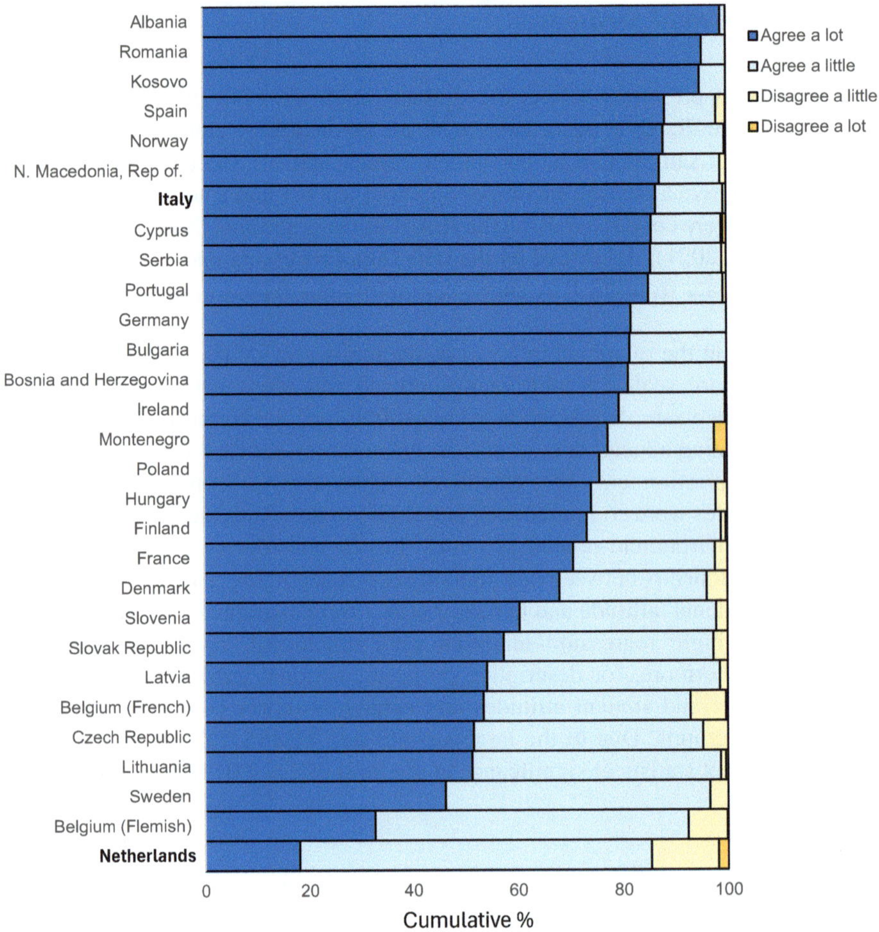

Fig. 5.1 Percentage of students whose teachers consider teaching environmental sustainability in schools a priority. Note: Data are from S6

of students whose teachers only agree a little with prioritizing sustainability education ranges from one percent of the total (Albania) to 67.2 percent (the Netherlands), with a relatively even spread of education systems in between these extremes. Concerning the two case education systems, almost all Italian students (99.4%) have teachers who agree with prioritizing sustainability education, whereas in the Netherlands, 85.3 percent of teachers agree at least a little that sustainability education should be prioritized.

Next, the analysis considers how often teachers report incorporating sustainability education activities in their teaching, making use of data from S4 (Fig. 5.2). The different sustainability education activities are more often incorporated by Italian teachers than by Dutch teachers. However, teacher responses vary slightly across the four activities: Italy reports the highest levels for some items but somewhat

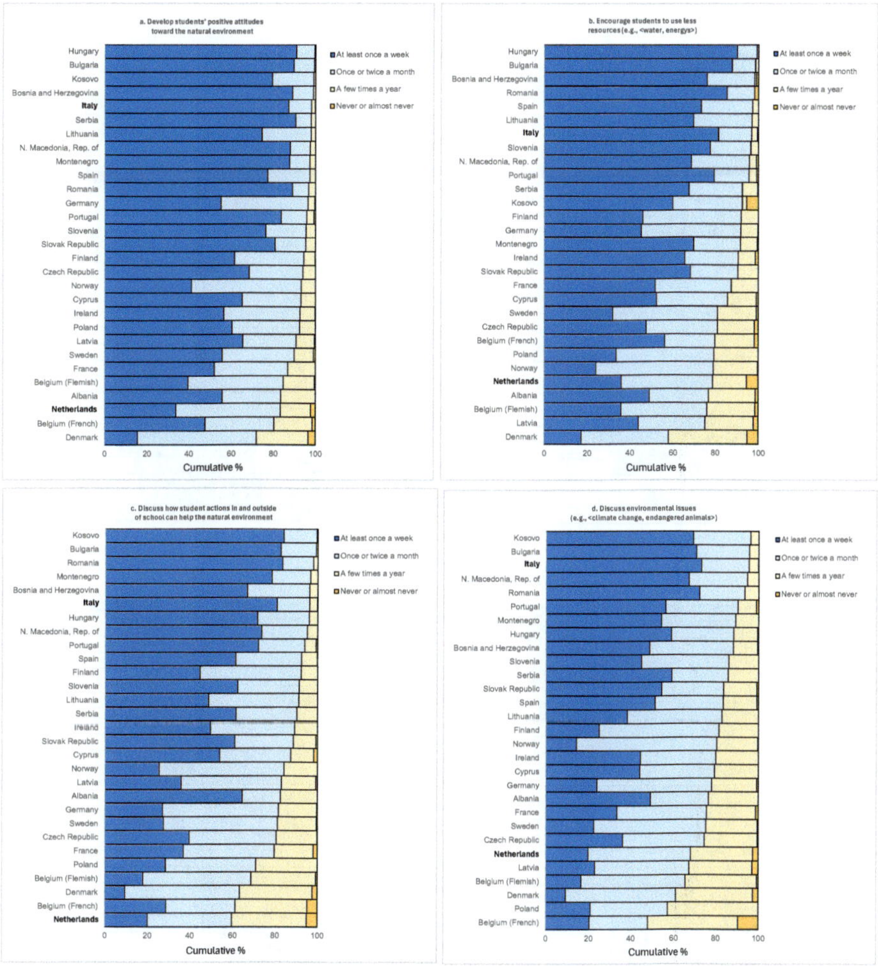

Fig. 5.2 Percentage of students whose teachers report on the question: "How often do you do this?". Note: Data is from S4

lower for others; conversely, the Netherlands reports the lowest levels for some items but somewhat higher levels for others.

The final question from the teacher questionnaire used to answer the first research question is S5, which concerns whether teachers report doing certain educational activities during sustainability education (Fig. 5.3). These activities range from taking students to natural areas to having students do research about environmental topics. Concerning the two case systems, Dutch students have teachers who report doing these activities less frequently than some colleagues in other European education systems. For some activities, Italian students have teachers who also report lower averages, but for other activities (for example, having students do research or

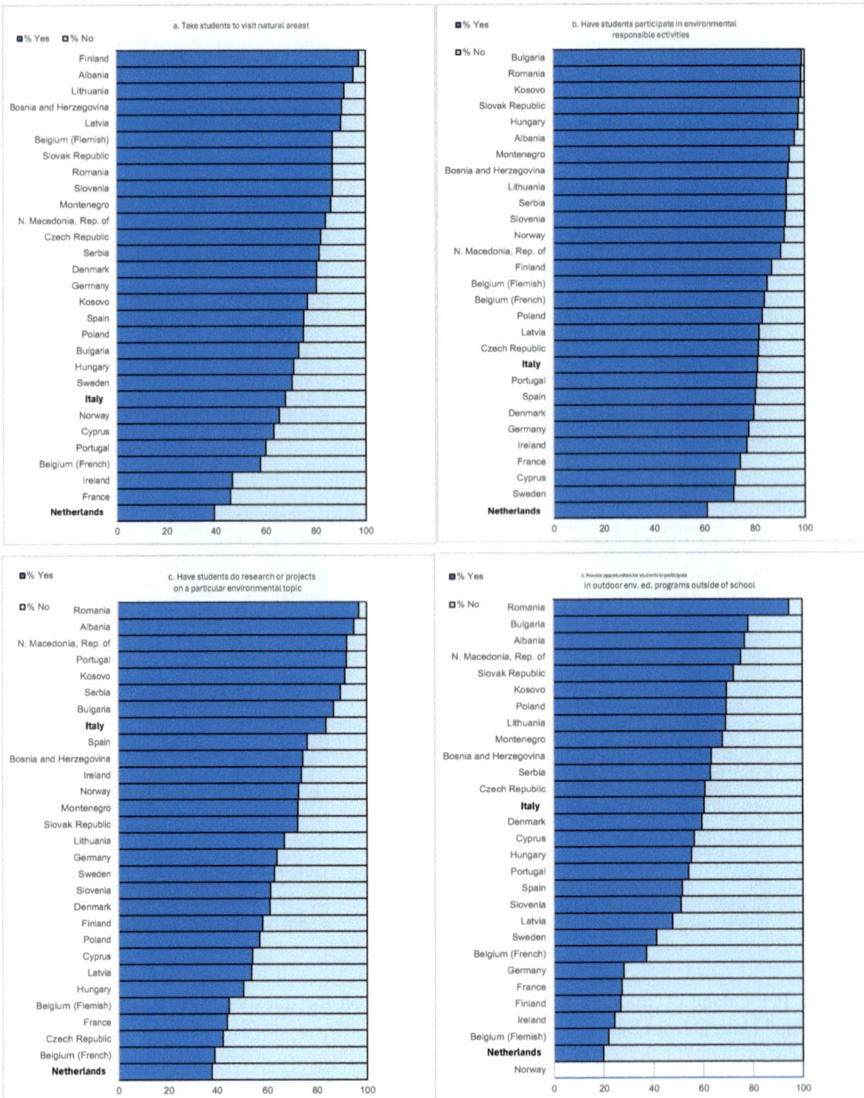

Fig. 5.3 Percentage of students whose teacher reports on the question: "Do you do this?". Notes: Data is from S5. For item d ("Provide opportunities for students to participate in environmental education programs outside of school"), no data were recorded in Norway

projects on a particular environmental topic; Fig. 5.3b), Italian teachers report doing this more often than most of their European colleagues.

Means for individual items in questions S4 and S5 indicate that Italy has higher percentages than the Netherlands in all instances (Table 5.2). For S4, a scale score was also calculated, but due to low internal reliability this was not done for the items of question S5. For the items in S4, the means approach the maximum, meaning

Table 5.2 Teacher answers on questions S4 and S5; means of teacher responses for individual items and overall scales for Italy, the Netherlands, and the selected education systems overall; including their association with the students' reported values and behaviors

	Education systems in focus				**Selected education systems**			
Teacher answers (weighted by students)	**Italy**		**Netherlands**		**Overall**		Relationship with students' values (α = 0.68)	Relationship with students' behaviors (α = 0.79)
Topics taught related to environment (S4)	Mean (SE)	SD (SE)	Mean (SE)	SD (SE)	Mean (SE)	SD (SE)	Mean (SE)	SD (SE)
Develop students' positive attitudes toward the natural environment (item S4-a)	2.85 (0.03)	0.42 (0.04)	2.15 (0.08)	0.75 (0.05)	2.60 (0.01)	0.56 (0.01)	**0.01 (0.00)**	0.01 (0.01)
Encourage students to use less resources (e.g., <water, energy>) (item S4-b)	2.78 (0.03)	0.51 (0.04)	2.09 (0.09)	0.86 (0.06)	2.45 (0.01)	0.66 (0.01)	**0.01 (0.00)**	0.01 (0.00)
Discuss how student actions in and outside of school can help the natural environment (item S4-c)	2.77 (0.03)	0.51 (0.04)	1.75 (0.09)	0.83 (0.04)	2.36 (0.01)	0.65 (0.01)	**0.01 (0.00)**	**0.01 (0.00)**
Discuss environmental issues (e.g., <climate change, endangered species>) (item S4-d)	2.68 (0.03)	0.56 (0.03)	1.85 (0.09)	0.76 (0.04)	2.21 (0.01)	0.70 (0.01)	**0.02 (0.00)**	**0.01 (0.00)**
Scale of *Topics taught related to environment* (α = 0.84)	*11.26 (0.09)*	*1.53 (0.09)*	*8.51 (0.25)*	*2.15 (0.15)*	*9.97 (0.03)*	*1.77 (0.02)*	***0.02 (0.00)***	***0.02 (0.00)***

(continued)

Table 5.2 (continued)

	Education systems in focus				**Selected education systems**			
Teacher answers (weighted by students)	**Italy**		**Netherlands**		**Overall**		Relationship with students' values ($\alpha = 0.68$)	Relationship with students' behaviors ($\alpha = 0.79$)
Activities related to environment (S5)	Yes (SE)	SD (SE)	Yes (SE)	SD (SE)	Yes (SE)	SD (SE)	Yes (SE)	SD (SE)
Take students to visit natural areas (e.g., <a pond or meadow>) (item S5-a)	0.68 (0.03)	0.47 (0.01)	0.39 (0.05)	0.49 (0.01)	0.76 (0.01)	0.39 (0.00)	**0.02 (0.00)**	**0.02 (0.00)**
Have students participate in environmentally responsible activities (e.g., <pick up trash>) (item S4-b)	0.80 (0.03)	0.40 (0.02)	0.62 (0.05)	0.49 (0.01)	0.86 (0.00)	0.32 (0.01)	**0.02 (0.00)**	**0.01 (0.00)**
Have students do research or projects on a particular environmental topic (e.g., <pollution, climate change>) (item S5-c)	0.83 (0.02)	0.37 (0.02)	0.38 (0.05)	0.49 (0.01)	0.68 (0.01)	0.42 (0.00)	**0.02 (0.00)**	**0.02 (0.01)**
Provide opportunities for students to participate in outdoor environmental education programs outside of school) (item S5-d)	0.59 (0.03)	0.49 (0.01)	0.2 (0.04)	0.40 (0.03)	0.54 (0.01)	0.46 (0.00)	**0.02 (0.00)**	**0.02 (0.01)**
Sum of *Activities related to environment*	*2.93 (0.07)*	*1.09 (0.04)*	*1.58 (0.12)*	*1.13 (0.06)*	*2.82 (0.01)*	*0.99 (0.01)*	***0.03 (0.00)***	***0.03 (0.00)***

The difference between Italy and the Netherland are significant for all items and scale scores
() standard errors in parenthesis
Standard errors <0.01 are reported as (0.00)

that, on average, Italian students have teachers who perform the discussed activities at least once a week. The Dutch averages, in contrast, more closely resemble once or twice a month. In all cases except S5-d ("Provide opportunities for students to participate in outdoor environmental education programs outside of school"), the standard deviation of the mean in the Netherlands is higher than that of Italy or of the selected education systems overall, reflecting greater variation in teacher responses within the Netherlands (Table 5.2).

The second research question concerns the association between the extent to which teachers encourage environmentally responsible behavior among students, and the environmental values and behaviors of their students. Concerning the associations between the teacher items of S4 and S5 with student values and behaviors, almost all of the slopes for individual items were significant, albeit negligible (see Cohen, 1988) (Table 5.2). These associations were calculated while retaining information at education system level. This means that with these data no association could be found for the educational activities that teachers employ and student environmental values and behaviors.

Next, the analysis explored the significance of the associations between teachers' responses to the constructed scale for environmental topics taught (S4) and student pro-environmental attitudes and behaviors. To review relationships between teacher topics taught and student environmental attitudes and behaviors, tertiles for the teacher topic taught index were computed and student environmental attitudes and behaviors averages were then compared across these tertile groups.

For student pro-environmental attitudes, in all but two education systems (Finland and Spain), no significant association could be found with teacher self-reported environmental topics taught (Fig. 5.4a). A similar pattern was found for the association between teacher responses to the constructed topics taught scale (S4) and student behavior (Fig. 5.4b). For one education system (Romania), a significant and positive association was found, whereas all other associations were not significant.

5.4 Discussion

In line with the sustainability education research community and international bodies such as the United Nations (United Nations, 2015) and the European Union (European Commission, Directorate-General for Education, Youth, Sport and Culture, 2022), international large-scale assessments such as TIMSS acknowledge the value of sustainability education to foster student empowerment for taking action towards a more sustainable future. Even though TIMSS was not initially designed to tap into sustainability education implementation topics, it now includes a focus on relevant environmental knowledge since the introduction of its Environmental Awareness Framework in 2019. The 2023 cycle reinforced this focus and added the Environmental Attitudes and Behaviors Framework (Reynolds & Komakhidze, 2022). This growing focus on environmental knowledge, attitudes,

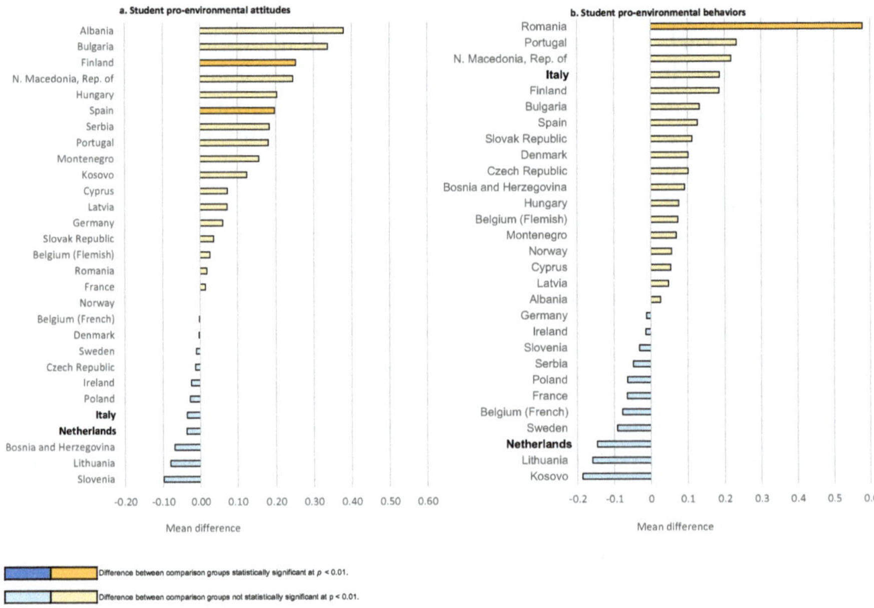

Fig. 5.4 Associations between teacher responses to the environmental topics taught scale (S4) and student pro-environmental (**a**) attitudes and (**b**) behaviors. Note: Dark shading marks significant associations

and behaviors resonates with the sense of urgency for climate action induced by increasingly undeniable effects of climate disruption on human life conditions (Intergovernmental Panel on Climate Change, 2023). Thus, TIMSS is answering calls for providing high-quality comparative data that enable education systems to monitor and evaluate the effects of educational strategies at a cross-national scale (Buckler & Creech, 2014), similar to ongoing efforts in other international large-scale assessments such as PISA and ICCS.[6]

The current chapter aims to support TIMSS in this endeavor by critically examining the 2023 cycle's alignment with relevant insights from recent empirical research findings regarding the potential for action-oriented sustainability education to foster students' competence to take action for the environment. Therefore, the overarching question guiding this chapter is whether the TIMSS data, which are high-quality from a statistical and measurement point of view, can help measure what is relevant for enabling education systems to monitor the environmental attitudes and behaviors of students and evaluate teachers' educational approaches for empowering students to contribute to building a sustainable future.

In this final section of the chapter, the research questions are revisited and addressed in light of the results of the analyses. Throughout, the selected variables'

[6]PISA = *Programme for International Student Assessment*; ICCS = International Civic and Citizenship Education Study.

alignment with the definitions of action (Jensen & Schnack, 2006; Mogensen et al., 2009; Sass et al., 2020) and an action-oriented educational approach (Sinakou et al., 2019; Varela-Losada et al., 2016) is examined. The chapter concludes with some recommendations to further strengthen future TIMSS cycles for monitoring and evaluation of sustainability education.

5.4.1 Educational Activities Reported by Teachers

The analyses show that, for all of the European education systems included in this volume, a vast majority of teachers report that, in their opinion, teaching environmental sustainability should be a priority in schools. Differences between education systems are slight (ranging from 85.3% in the Netherlands to 100% in Albania; Fig. 5.1) and may be affected by differences in how strongly teachers express opinions (Zaller & Feldman, 1992) or tied to actual differences in the perceived priority that teachers ascribe to this issue. Regardless of possible nuances in the extent to which they agreed, the findings suggest there is a strong consensus on the importance of sustainability education among the participants in TIMSS. There is great support for prioritizing sustainability education across all European education systems selected for this volume. It should be noted that teachers express their support for prioritizing sustainability "at school," which does not necessarily imply that they think this is a priority for science education. They may feel that it is the school's broader responsibility to address sustainability throughout the curriculum or even through a whole school approach (Holst, 2023). The current operationalization of question S6 does not allow differentiation between these diverse interpretations.

When examining the activities teachers report actually implementing in their (science) education, differences between education systems become more evident. This raises questions regarding how strong teacher support for prioritizing environmental sustainability manifests itself in classroom practices. Given the nature of the data used in these analyses, the causes of such differences cannot be determined with certainty and can only be speculated upon. Possible explanations include whether teachers and schools have sufficient freedom, resources, opportunities for professional development and learning in teacher or school communities, and time to implement relevant class practices (Rushton & Walshe, 2025; Taylor et al., 2019). Reports such as those by Eurydice (*Learning for sustainability in Europe: Building competences and supporting teachers and schools*) highlight the great diversity in top-level curricula for integrating environmental sustainability across education systems (European Commission/EACEA/Eurydice, 2024). Considering that, based on the Eurydice analyses, Italy focuses more strongly on sustainability education across different subjects and even in a cross-curricular fashion, whereas the Netherlands hardly incorporates sustainability in the curriculum at all, these different teacher responses may be at least partially explained by the place sustainability education and sustainability competences hold in top-level curricula.

An additional consideration is whether the educational activities included in the TIMSS 2023 cycle enable measuring action-oriented sustainability education implementation, or whether aspects of action-oriented sustainability education remain uncovered by the current TIMSS questionnaire. Teachers might also employ other practices in their teaching that align just as well or even better with the definition of an action-oriented educational approach as defined in sustainability education research. This thought will be revisited later in this discussion section.

When analyzing how often and which environmental sustainability activities are included by teachers in their science education, interpretation of the results becomes less straightforward. The results show that, depending on the particular activity, education systems rank higher or lower in comparison to each other. The questionnaire items regarding teaching activities included the frequency (item S4 in the teacher questionnaire; see Fig. 5.2 and Table 5.2) of developing students' positive attitudes towards the natural environment (the item with the highest mean value overall across all included education systems) or discussing environmental issues such as climate change and endangered animals (the item with the lowest mean value overall across all included education systems). Another question assessed whether teachers implemented activities (question S5 in the teacher questionnaire; see Fig. 5.3 and Table 5.2) such as having students participate in environmentally responsible activities (for example, picking up trash) (the item with the highest mean score overall) or providing opportunities for students to participate in outdoor environmental education programs outside school (the item with lowest mean score overall). Focusing on Italy and the Netherlands, the sorted data show Italy ranking between eighth and first position across S4 items regarding frequency (Fig. 5.2), and the Netherlands ranging from second-to-last position for frequency of developing students' positive attitudes to sixth-to-last position for frequency of encouraging students to use less resources (for example, water or energy).

These examples illustrate that no clear pattern emerges in the different education systems' relative scores across the different activities, suggesting that the activities are too diverse to form an overall scale. The unsatisfactorily low Cronbach's alpha value for a potential scale that would group the S5 items, supports this interpretation. The items in S5 are not aligned in such a way as to assume they jointly measure a single construct; in other words, they do not have a convincingly high internal validity. Looking at the individual items (Table 5.1)—which concern diverse activities such as visiting natural areas, having students perform environmentally friendly activities, and having students do research on environmental topics—illustrates why this might be the case. Moreover, there may be a discrepancy between what teachers report doing and how students actually experience what happens at school. Therefore, it would be valuable to also include students' perspectives on what educational approaches are implemented in the science classroom (Isac et al., 2022; Sass et al., 2023; Treviño & Carrasco, 2025). Additionally, a greater focus on scale design, including a minimum of three conceptually related items per scale to avoid under-identification of the subconstruct (Brown, 2015), would further strengthen TIMSS data. Future TIMSS cycles may find inspiration for newly validated scales in work by Sass et al. (2023, 2024) and, which directly address the topic of action-oriented sustainability education.

When reflecting on the alignment of the current TIMSS items with the framework of action-oriented sustainability education, a final consideration arises. The current items tap into activities selected by the teacher rather than actions regarding wicked problems that are co-decided on by students (see, for example, item S4-c, Table 5.1). Action-oriented sustainability education refers to an educational approach that encourages students to engage in actions (learning by doing) focused on finding solutions to real-world controversial problems (Sinakou et al., 2019; Varela-Losada et al., 2016). Previous empirical work shows that openness in educational activities appears to foster students' action competence (Sass et al., 2023). Current studies provide examples of how to write items that emphasize the voluntary character of action and participatory character of action-oriented sustainability education, resulting in potential items such as, "At our school, we decide together with the teacher what we need to learn in order to contribute to solving a local climate problem" (after Sass et al., 2024). Similarly, regarding the need to embed sustainability education in the local community for it to be consistent with action-oriented teaching, scales might include items such as the extent to which students agree their school encourages them to "organize an action in their local community for a good life for everyone without damaging the planet" or to "collaborate on pro-environmental actions" (after Sass et al., 2023).

5.4.2 Associations Between Reported Teacher Activities and Attitudes and Behaviors of Students

When comparing education systems, the results suggest no associations between the frequency nor the type of educational activities that teachers include in their teaching on the one hand, and student environmental attitudes and behaviors on the other. Slopes for all individual items in S4 and S5 were negligible for both the case systems of Italy and the Netherlands and for the aggregated overall mean of the selected education systems. While just a single positive and significant association could be found between environmental topics taught (S4) and student pro-environmental behaviors (for Romania, see Fig. 5.4b), this association was nonsignificant for all other education systems. Similarly, the number of environmental topics taught (S4) was positively associated with student environmental attitudes in two education systems (Finland and Spain, see Fig. 5.4a), whereas it was nonsignificant in all other selected education systems. Even though the 2024 Eurydice report shows that the Italian curriculum has a strong sustainability competence focus, and teachers report doing sustainability education activities in their lessons more often (S4/S5), no significant association could be found with environmental attitudes or behaviors of the students in Italy. The same lack of significant associations between teaching activities and student outcomes was found in the Netherlands, where the top-level curriculum does not contain a strong focus on sustainability competences. In this respect it might be interesting, in follow up studies, to explore explanations for the positive association between teacher activities and student environmental attitudes or behaviors in Finland, Spain, or Romania, which are moderately positive (as shown in Fig. 5.4). For example, is that positive relationship associated with teaching methods (such as student-directed activities or community-focused education) or the requirement to teach specific environmental topics?

The overall findings contrast with empirical research findings described in the literature which suggest an action-oriented teaching approach helps foster students' action competence (Sass et al., 2023; Torsdottir et al., 2024). This discrepancy warrants closer scrutiny regarding the extent to which the proposed teaching activities in TIMSS 2023 align with action-oriented educational approaches as defined in the research literature. First, attention is directed to the alignment of TIMSS S5 items concerning teachers' implementation of certain educational activities within the context of environmental sustainability in science education. The empirical findings presented in the results section (Fig. 5.3) revealed that these items focus on activities rather than actions. Taking students to visit natural areas, providing opportunities for students to participate in outdoor environmental education programs outside of school, having students participate in environmentally responsible activities, and having students do research or projects on a particular environmental topic all imply choices made by the teacher rather than the students. Consequently, these are *activities*, rather than voluntary student behaviors. Therefore, they cannot be considered as opportunities for student *action* or as supporting students to develop their *action competence*. The phrasing "have students do" and "have students participate in" certain activities reflects a strongly teacher-directed approach, whereas action, and hence also action-oriented teaching, necessitates recognition of and opportunities for student-led initiative and co-decision-making. Replacing, for example, "Have students participate in environmentally responsible activities" with "Provide students with opportunities to participate in environmentally responsible activities" might move these items closer to the notion of action-oriented teaching.

Drawing on the principles of action-orientation, in addition to offering students opportunities to make choices collaboratively with peers and jointly with the teacher, activities should also reference contributing to solving wicked problems rather than being limited to activities such as taking students to visit natural areas. Revisiting the teacher-led activities and student behavior scales to align them more closely with the notions of teachers' action-oriented educational approaches and student action, respectively, could enhance the informative value of future TIMSS cycles for education systems' monitoring and evaluation efforts. Empirically validated instruments such as the Self-Perceived Action Competence for Sustainability questionnaire (SPACS; Olsson et al., 2020) and the Action Competence in Sustainable Development Questionnaire (ACiSD-Q; Sass et al., 2020; van Harskamp et al., 2024) may provide useful models, as they address voluntary student behavior aimed at contributing to a more sustainable future by tackling real-world problems—reflecting *action*—through learning by doing both within and beyond the school context—reflecting *action-oriented* educational approaches (Sinakou et al., 2019; Varela-Losada et al., 2016).

5.5 Conclusions

One of the main contributions of this chapter is that it reveals extensive support for sustainability education among teachers in all the TIMSS education systems included in the current volume. The analyses further demonstrated that this support is not consistently translated into concrete, impactful sustainability education, and that the TIMSS operationalization could be refined to better align with current understandings of action, action competence, and action-oriented teaching and learning.

Considering all results presented in this chapter, a concluding reflection is that the findings do not fully resonate with empirical evidence in the broader field of sustainability education. This may be due to the fact that, while TIMSS gathers data on the frequency of educational activities, it does not provide insight into the quality of those educational activities: more does not necessarily mean more effective. For example, a teacher may discuss the relevance of environmental protection multiple times per month or per week, but if this is not done effectively, it is unlikely to produce meaningful learning outcomes, such as increased pro-environmental action taking, attitudes, or behaviors among students. Future TIMSS cycles would benefit from incorporating measures that capture both quality and quantity of sustainability education. Such an approach would align more closely with the progression from Vision 1 (understanding science), through Vision 2 (scientific literacy and application in daily life), toward Vision 3 for science education (science for transformation) (Sjöström & Eilks, 2018).

In summary, TIMSS provides statistically high-quality data that can support education systems in monitoring the implementation of action-oriented sustainability education. The utility of TIMSS data would be further enhanced if additional scales were more closely aligned with the existing body of research literature on action (complementing the current TIMSS scales of environmental attitudes), action competence, and action-oriented sustainability education approaches in the science classroom (Bianchi et al., 2022; UNESCO, 2020). This chapter closely examined the sustainability education teaching variables available in the TIMSS 2023 data, and a similar exercise for a closer examination of student action in the TIMSS 2023 behavior scale is recommended. Adding items that emphasize action-oriented teaching and learning to future TIMSS questionnaires, as discussed above, would enable education systems to monitor both teacher practices and the development of students' action competence. Such information would provide education systems with high-quality, reliable and valid data, thereby leveraging TIMSS to support the effective implementation of sustainability education policies and practices.

Funding This study is part of the IMP>ACT project, which develops an impact assessment framework for sustainability and climate change education. IMP>ACT is funded by the European Union though Grant agreement ID 101137351. Views and opinions expressed are those of the author(s) only and do not necessarily reflect those of the European Union or CINEA. Neither the European Union nor the granting authority can be held responsible for them.

References

Bianchi, G., Pisiotis, U., & Cabrera Giraldez, M. (2022). *GreenComp: The European sustainability competence framework*. European Commission: Joint Research Centre. Publications Office of the European Union. https://doi.org/10.2760/13286

Boeve-de Pauw, J., Gericke, N., Olsson, D., & Berglund, T. (2015). The effectiveness of education for sustainable development. *Sustainability, 7*(11), 15693–15717. https://doi.org/10.3390/su71115693

Boeve-de Pauw, J., & Van Petegem, P. (2010). A cross-national perspective on youth environmental attitudes. *The Environmentalist, 30*, 133–144. https://link.springer.com/article/10.1007/s10669-009-9253-1

Boeve-de Pauw, J., & Van Petegem, P. (2018). Eco-school evaluation beyond labels: The impact of environmental policy, didactics, and nature at school on student outcomes. *Environmental Education Research, 24*(9), 1250–1267. https://doi.org/10.1080/13504622.2017.1307327

Bogner, F. X., & Wiseman, M. (1999). Toward measuring adolescent environmental perception. *European Psychologist, 4*(3), 139–151. https://doi.org/10.1027/1016-9040.4.3.139

Bogner, F. X., & Wiseman, M. (2006). Adolescents' attitudes towards nature and environment: Quantifying the 2-MEV model. *Environmentalist, 26*, 247–254. https://doi.org/10.1007/s10669-006-8660-9

Brown, T. A. (2015). *Confirmatory factor analysis for applied research* (Methodology in the social sciences) (2nd ed.). Guilford Press.

Buckler, C., & Creech, H. (2014). *Shaping the future we want. UN Decade of Education for Sustainable Development (2005–2014). Final report.* UNESCO Publishing. https://unesdoc.unesco.org/ark:/48223/pf0000230171

Coertjens, L., Boeve-de Pauw, J., De Maeyer, S., & Van Petegem, P. (2010). Do schools make a difference in their students' environmental attitudes and awareness? Evidence from PISA 2006. *International Journal of Science and Mathematics Education, 8*(3), 497–522. https://doi.org/10.1007/s10763-010-9200-0

Cohen, J. (1988). *Statistical power analysis for the behavioral sciences.* Lawrence Erlbaum Associates.

European Commission, Directorate-General for Education, Youth, Sport and Culture. (2022). Proposal for a Council Recommendation on learning for environmental sustainability. COM/2022/11 final. https://eur-lex.europa.eu/legal-content/EN/TXT/?uri=CELEX:52022DC0011

European Commission/EACEA/Eurydice. (2024). *Learning for sustainability in Europe: Building competences and supporting teachers and schools. Eurydice report.* Publications Office of the European Union. https://eurydice.eacea.ec.europa.eu/publications/learning-sustainability-europe-building-competences-and-supporting-teachers-and

Gong, B., & Zheng, Y. (2021). More is not always better: A study of country-level factors associated with adolescents' environmental attitudes using a multilevel analysis of PISA 2006. *Educational Policy Analysis Archives, 29*, 125. https://doi.org/10.14507/epaa.29.4846

Hasslöf, H., & Malmberg, C. (2015). Critical thinking as room for subjectification in education for sustainable development. *Environmental Education Research, 21*(2), 239–255. https://doi.org/10.1080/13504622.2014.940854

Holst, J. (2023). Towards coherence on sustainability in education: A systematic review of Whole Institution Approaches. *Sustainability Science, 18*, 1015–1030. https://doi.org/10.1007/s11625-022-01226-8

IEA. (2021). *Data and tools. Tools.* https://www.iea.nl/data-tools/tools

Intergovernmental Panel on Climate Change. (2023). *Climate change 2023: Synthesis report.* Contribution of Working Groups I, II and III to the Sixth Assessment Report of the Intergovernmental Panel on Climate Change [Core Writing Team, H. Lee & J. Romero (Eds.)]. https://doi.org/10.59327/IPCC/AR6-9789291691647.

Isac, M. M., Sandoval-Hernández, A., & Sass, W. (Eds.). (2025). *Knowledge and willingness to act pro-environmentally. Perspectives from IEA TIMSS 2019 and ICCS 2016 data* (IEA research for education) (Vol. 16). Springer. https://link.springer.com/book/10.1007/978-3-031-76033-4

Isac, M. M., Sass, W., Boeve-de Pauw, J., De Maeyer, S., Schelfhout, W., Van Petegem, P., & Claes, E. (2022). Differences in teachers' professional action competence in education for sustainable development: The importance of teacher co-learning. *Sustainability, 14*(2), 767. https://doi.org/10.3390/su14020767

Jensen, B. B. (2000). Health knowledge and health education in the democratic health-promoting school. *Health Education, 100*(4), 146–154. https://doi.org/10.1108/09654280010330900

Jensen, B. B., & Schnack, K. (2006). The action competence approach in environmental education. Reprinted from *Environmental Education Research, (1997) 3*(2), 163–178. *Environmental Education Research, 12*(3–4), 471–486. https://doi.org/10.1080/13504620600943053.

Knippels, M. C. P. J., & van Harskamp, M. (2018). An educational sequence for implementing socio-scientific inquiry-based learning. *School Science Review, 100*(371), 46–52.

Krnel, D., & Naglic, S. (2009). Environmental literacy comparison between ECO-schools and ordinary schools in Slovenia. *Science Education International, 20*, 5–24.

Lantz, B. (2013). The large sample size fallacy. *Scandinavian Journal of Caring Sciences, 27*(2), 487–492. https://doi.org/10.1111/j.1471-6712.2012.01052.x

List, M. K., Schmidt, F. T. C., Mundt, D., & Föste-Eggers, D. (2020). Still green at fifteen? Investigating environmental awareness of the PISA 2015 population: Cross-national differences and correlates. *Sustainability, 12*(7), 2985. https://doi.org/10.3390/su12072985

Macintyre, T., Tilbury, D., & Wals, A. (2024). *Education and learning for sustainable futures: 50 years of learning for environment and change*. Taylor & Francis. https://doi.org/10.4324/9781003467007

Mogensen, F., Breitng, S., Hedegaard, K., Nielsen, K., & Schnack, K. (2009). *Action competence, conflicting interests and environmental education: The MUVIN Programme*. Danmarks Pædagogiske Universitetsforlag. https://www.ucviden.dk/en/publications/action-competence-conflicting-interests-and-environmental-educati

Muthén, L. K., & Muthén, B. O. (2017). *Mplus user's guide* (8th ed.). Muthén & Muthén.

Olsson, D., Gericke, N., Sass, W., & Boeve-de Pauw, J. (2020). Self-perceived action competence for sustainability: The theoretical grounding and empirical validation of a novel research instrument. *Environmental Education Research, 26*(5), 742–760. https://doi.org/10.1080/13504622.2020.1736991

Reynolds, K. A., & Komakhidze, M. (2022). *TIMSS 2023 environmental attitudes and behaviors framework*. TIMSS & PIRLS International Study Center, Boston College. https://timssandpirls.bc.edu/timss2023/frameworks/pdf/T23-Environmental-Attitudes-and-Behaviors-Framework.pdf

Roczen, N., Kaiser, F., Bogner, F., & Wilson, M. (2014). A competence model for environmental education. *Environment and Behavior, 46*(8), 972–992. https://doi.org/10.1177/0013916513492416

Rushton, E. A. C., & Walshe, N. (2025). Curriculum making and climate change and sustainability education: A case study of school teachers' practices from England, UK. *Environmental Education Research, 31*(5), 1083–1097. https://doi.org/10.1080/13504622.2025.2471990

Sass, W., Boeve-de Pauw, J., Olsson, D., Gericke, N., De Maeyer, S., & Van Petegem, P. (2020). Redefining action competence: The case of sustainable development. *The Journal of Environmental Education, 51*(4), 292–305. https://doi.org/10.1080/00958964.2020.1765132

Sass, W., Boeve-de Pauw, J., Olsson, D., van der Werf, L., & Gericke, N. (2024). Student perspectives on quality education at their school: Toward a valid and reliable questionnaire to monitor quality education for sustainability teaching (QUEST-Q). *Environmental Education Research*, 1–28. https://doi.org/10.1080/13504622.2025.2553178

Sass, W., De Maeyer, S., Boeve-de Pauw, J., & Van Petegem, P. (2023). Effectiveness of education for sustainability: The importance of an action-oriented approach. *Environmental Education Research, 30*(4), 479–498. https://doi.org/10.1080/13504622.2023.2229543

Sinakou, E., Donche, V., Boeve-de Pauw, J., & Van Petegem, P. (2019). Designing powerful learning environments in education for sustainable development: A conceptual framework. *Sustainability, 11*(21), 5994. https://doi.org/10.3390/su11215994

Sjöström, J., & Eilks, I. (2018). Reconsidering different visions of scientific literacy and science education based on the concept of building. In Y. J. Dori, Z. R. Mevarech, & D. R. Baker (Eds.), *Cognition, metacognition, and culture in STEM education* (Innovations in science education and technology) (Vol. 24, pp. 65–88). Springer. https://doi.org/10.1007/978-3-319-66659-4_4

Taylor, N., Quinn, F., Jenkins, K., Miller-Brown, H., Rizk, N., Prodromou, T., Serow, P., & Taylor, S. (2019). Education for sustainability in the secondary sector—A review. *Journal of Education for Sustainable Development, 13*(1), 102–122. https://doi.org/10.1177/0973408219846675

Torsdottir, A. E., Olsson, D., & Tonette, A. (2024). Developing action competence for sustainability – Do school experiences in influencing society matter? *Global Environmental Change, 86*, 102840. https://doi.org/10.1016/j.gloenvcha.2024.102840

Treviño, E., & Carrasco, D. (2025). School factors and students' willingness to act pro-environmentally in the future. In M. M. Isac, A. Sandoval-Hernández, & W. Sass (Eds.), *Knowledge and willingness to act pro-environmentally. Perspectives from IEA TIMSS 2019 and ICCS 2016 data* (IEA research for education) (Vol. 16, pp. 151–167). Springer. https://doi.org/10.1007/978-3-031-76033-4_10

UNESCO. (2020). *Education for sustainable development: A roadmap*. UNESCO Publishing. https://unesdoc.unesco.org/ark:/48223/pf0000374802

United Nations. (2015). *Transforming our world: The 2030 Agenda for Sustainable Development* (A/RES/70/1). https://sdgs.un.org/2030agenda

van Harskamp, M., De Maeyer, S., Sass, W., Van Petegem, P., & Boeve-de Pauw, J. (2024). Measurement invariance of the action competence in sustainable development questionnaire: Can we compare between groups? *Environmental Education Research, 31*(3), 498–511. https://doi.org/10.1080/13504622.2024.2382967

Varela-Losada, M., Vega-Marcote, P., Pérez-Rodríguez, U., & Álvarez-Lires, M. (2016). Going to action? A literature review on educational proposals in formal environmental education. *Environmental Education Research, 22*(3), 390–421. https://doi.org/10.1080/13504622.2015.1101751

von Davier, M., Fishbein, B., & Kennedy, A. (Eds.). (2024). *TIMSS 2023 technical report (methods and procedures)*. TIMSS & PIRLS International Study Center, Boston College. https://timss2023.org/methods

Wiseman, M., & Bogner, F. X. (2003). A higher-order model of ecological values and its relationship to personality. *Personality and Individual Differences, 34*(5), 783–794. https://doi.org/10.1016/S0191-8869(02)00071-5

Yeung, S. P.-M. (2004). Teaching approaches in geography and students' environmental attitudes. *Environmentalist, 24*, 101–117. https://link.springer.com/article/10.1007/s10669-004-4801-1

Yin, L., & Reynolds, K. A. (2024). Creating and interpreting the TIMSS context questionnaire scales. In M. von Davier, B. Fishbein, & A. Kennedy (Eds.), *TIMSS 2023 technical report (methods and procedures)* (pp. 15.1–15.14). TIMSS & PIRLS International Study Center, Boston College. https://doi.org/10.6017/lse.tpisc.timss.rs2550

Zaller, J., & Feldman, S. (1992). A simple theory of the survey response: Answering questions versus revealing preferences. *American Journal of Political Science, 36*(3), 579–616. https://doi.org/10.2307/2111583

Jelle Boeve-de Pauw is an Associate Professor of Science Education at the Freudenthal Institute, part of the Faculty of Science at Utrecht University. Trained as both a biologist and an educational scientist, his research focuses on sustainability education, science education, and citizenship education, with particular interest in how these domains intersect. He earned his PhD in 2011 from the

University of Antwerp (Belgium) and received the 2014 Rudi Verheyen Early Career Award for outstanding research with relevance to environmental policy in Flanders. His work emphasizes issues of measurement and evaluation in education. His most cited publications explore the educational effectiveness of Eco-Schools and similar programs, the use of international large-scale assessment data to evaluate environmental education, and conceptual frameworks for understanding action-taking and action competence as educational outcomes. At Utrecht University, he teaches in the Master's program Science Education and Communication, contributing to courses on research methodology, science-society relations, and the didactics of sustainability.

Michiel van Harskamp is a postdoctoral researcher in science education with a focus on sustainability education at the Freudenthal Institute, Utrecht University, the Netherlands. He has a background in Biology, and a PhD in fostering Environmental Citizenship through science education (2023). In his PhD, Van Harskamp focused on Socio-Scientific Issues related to sustainability and Inquiry-Based Science Education, and their potential to promote Environmental Citizenship through Socio-Scientific Inquiry-Based Learning. For this, he collaborated with a team of six science teachers in a Lesson Study team. Through several Lesson Study cycles, the team researched both teacher and student EC competence development. He currently works for the Horizon-Europe IMP>ACT project, a research project where he researches assessment of action competence as a learning outcome of sustainability and climate change education. Furthermore, he researches climate change emotions of young people in collaboration with colleagues from Utrecht University's Faculty of Social and Behavioural Sciences.

Wanda Sass is a teacher in secondary education. Next to this, Wanda is a postdoctoral researcher at the Freudenthal institute (Utrecht University) within the HARP-project (*Hope and Resilience Perspectives in Climate Change Education*). Her current research focuses on emotion-sensitive climate change and sustainability education, and action competence as one of its desired learning outcomes. She was involved in research into nature-connectedness, motivation towards the environment, and a whole institution approach in a context of implementing education for sustainability in compulsory and higher education. In collaboration with Maria Magdalena Isac and Andres Sandoval-Hernandez, she recently co-edited *Knowledge and Willingness to Act Pro-Environmentally: Perspectives from IEA TIMSS 2019 and ICCS 2016 Data*, in the IEA Research for Education series. In 2022, she obtained a doctoral degree in educational sciences with a dissertation on action competence (University of Antwerp). She loves bridging the gap between educational research and practice, learning from both perspectives.

Elisa Caponera PhD, is a senior researcher at INVALSI. She earned a PhD in Innovation and Evaluation of Education Systems from Roma Tre University and a specialization in Psychological Evaluation from Sapienza University of Rome. She has expertise in developing and validating questionnaires and serves as Italy's National Research Coordinator for IEA projects TIMSS and ICILS. Her research focuses on parental involvement and factors influencing learning outcomes.

Riccardo Pietracci is a Research Associate at INVALSI, where he serves as national contact for computer-based assessments in international large-scale studies. His research focuses on civic and citizenship education, sustainability and climate change education, and the assessment of complex constructs through innovative digital test environments.

Laura Palmerio holds a degree in Psychology, a PhD in Innovation and Evaluation of Education Systems, and a specialization in Health Psychology. Senior Researcher at INVALSI, she leads the International Surveys unit and coordinates Italy's participation in IEA's TIMSS, PIRLS, ICCS, and OECD's PISA and TALIS. Her research focuses on equity in education, the relationship between reading and mathematics literacy, civic and citizenship education, and information literacy.

Chapter 6
The Role of School Organizations in Promoting Sustainability

Inés García-Bohórquez, Dries Verhelst, Fernando Martínez-Abad, and Camilo Ruiz

6.1 Introduction

The climate crisis—alongside interconnected environmental challenges such as biodiversity loss, resource depletion, and pollution—poses profound risks to ecosystems, economies, and human well-being (Intergovernmental Panel on Climate Change, 2023). Achieving sustainability, and interconnected goals such as decarbonization of the economy, requires a fundamental transformation in how societies think, learn, and act in relation to the environment (European Commission, 2019). The growing urgency, together with the shifting attention away from these global environmental challenges due to an increased focus on defense, geopolitical tensions and economic instability, demonstrates the even larger need for education that enables students to take informed action for a sustainable future for all. In this context, education is an essential tool to achieve the goals of sustainability. It equips individuals and communities with the knowledge, values, and skills needed to navigate uncertainty, make informed decisions, build resilience, and collectively shape more sustainable futures (United Nations Educational, Scientific and Cultural Organization [UNESCO], 2022, 2024).

I. García-Bohórquez (✉) · C. Ruiz
EMC3 Research Group, Department of Science Education, University of Salamanca, Salamanca, Spain
e-mail: inesgarbohorquez@usal.es; camilo@usal.es

D. Verhelst
Faculty of Social Sciences, Edubron research group, University of Antwerp, Antwerp, Belgium
e-mail: Dries.verhelst@uantwerpen.be

F. Martínez-Abad
University Institute of Educational Sciences, University of Salamanca, Salamanca, Spain
e-mail: fma@usal.es

D. Hastedt et al. (eds.), *Educating Pupils for Environmental Sustainability in Europe*, IEA Research for Education 18,
https://doi.org/10.1007/978-3-032-14725-7_6

Global frameworks such as the Sustainable Development Goals (SDGs) (particularly SDGs 4.7 and 12.8), the European sustainability competence framework (GreenComp), and related concepts like climate and action competence emphasize the central role of education in equipping students with the knowledge, values, skills, and agency needed to address environmental and climate crises (Bianchi et al., 2022; United Nations, 2015). Among these, GreenComp stands out as a comprehensive framework that defines sustainability competence as a core competence for lifelong learning in the twenty-first century. It outlines four interconnected areas: *embodying sustainability values*, *embracing complexity*, *envisioning sustainable futures*, and *acting for sustainability*. Together, these areas reflect the diverse capabilities young people need to thrive and lead in a rapidly changing world (Bianchi et al., 2022; Wals et al., 2024).

While these frameworks provide a clear vision for sustainability education, the responsibility for translating these policy objectives into practice falls largely on schools. They are increasingly recognized not only as sites for learning but also as communities that can foster environmental responsibility and sustainable behaviors in children from a young age (White et al., 2023). Schools are not simply a vehicle for curricular content; they are complex and dynamic organizations whose structures, resources, and organization influence educational outcomes (Mogren et al., 2019; Scott, 2013). One example of the decisive role schools plays in advancing the goals of education for sustainable development (ESD) is the whole school approach (WSA), which offers a comprehensive model for embedding sustainability into the core functions of educational institutions. The WSA integrates sustainability into all dimensions of school life: leadership and governance, teaching and learning, community engagement, school operations, and the school's culture.

Research using international large-scale assessments, such as those conducted by the International Association for the Evaluation of Educational Achievement (IEA) and the Organisation for Economic Co-operation and Development (OECD), reinforces the importance of school-level factors, such as organizational structures, resource availability, and objectives, in shaping educational outcomes (Boeve-de Pauw et al., 2015; Klieme & Nilsen, 2022; List et al., 2020; Verhelst et al., 2022). More specifically, the organizational characteristics also determine whether educational institutions can effectively create and improve environmental competencies (Sass et al., 2020).

6.1.1 Study Contribution and Rationale

In recent decades, education for sustainability has gained prominence, becoming an explicit component of educational policy and curricula across Europe (UNESCO, 2017). Despite recent political backlash in some contexts, such as the withdrawal of the United States of America from the Paris agreement and the relaxation of decarbonization goals in Europe, the role of education in promoting sustainability, supporting environmental protection, and contributing to decarbonization remains widely recognized. However, empirical research exploring how schools shape student outcomes—particularly in primary education and across diverse national contexts—remains limited (Verhelst et al., 2023). Addressing this gap, this chapter advances the call for further research by conducting a secondary analysis of data from the 2023 Trends in International Mathematics and Science Study (TIMSS).

This study aims to identify practices at the level of school organization that could facilitate desirable outcomes of ESD, as well as barriers that may limit or prevent their implementation. The analysis focuses on how school characteristics linked to a WSA to sustainability education can facilitate the development of both environmental knowledge and attitudes. Using the ESD-effective school framework developed by Verhelst et al. (2020, 2024), attention is given to the influence of factors such as shared vision, adaptability, supportive relationships, and collective efficacy. This framework examines how factors such as academic emphasis, resource shortages, sense of belonging, and parental engagement are related to sustainability outcomes.

The results from this research can support schools in integrating sustainability into their daily operations and teaching practices, and provide evidence to inform policymakers, school leaders, and educational stakeholders. Strengthening these connections will help create supportive environments that foster not only academic success but also environmental responsibility and sustainability among students.

6.1.2 Research Questions

This research examines the relationships between school-level characteristics and students' environmental knowledge and attitudes. While some studies in this volume focus on the influence of educational policy (see Chap. 2) and the role of student engagement, the current analysis emphasizes the role of the schools themselves. Given that general frameworks like ESD and GreenComp are related to competences (including knowledge and attitudes), this study investigates the differences between these two factors across education systems.

Within the scope of the ESD-effective school framework, this chapter explores how structural and resource-related factors such as the availability of science materials, school size, and geographical location may influence the promotion of sustainability education within schools. Through this, the aim is to identify the organizational variables that most effectively foster a WSA to sustainability. In addressing these aims, the chapter is guided by the following research questions:

1. How do knowledge, attitudes and school emphasis on sustainability vary across education systems among fourth-grade students?
2. How do different variables within the school organization, as described in the ESD-effective school framework, relate to students' environmental knowledge?
3. How do different variables within the school organization, as described in the ESD-effective school framework, relate to students' environmental attitudes?

The analysis presented emphasizes the potential for school organizations to adapt to and advance the policy shifts towards sustainable education. By exploring how school-level factors are interlinked and connected to student outcomes, this chapter paves the path for a deeper understanding of how educational institutions can play a transformative role in empowering future generations to act sustainably and responsibly.

6.2 ESD-Effective School Framework

Competences related to sustainability (GreenComp, climate and action competences) highlight the need to go beyond knowledge about sustainability (Borgonovi et al., 2022). While knowledge is essential, research highlights that attitudes and emotions play a crucial role in fostering sustainable behaviors (Ballegeer et al., 2024). Schools must integrate these affective dimensions with active participation and values. A school's organizational environment, including its leadership, resources, and engagement with students and parents, shapes the extent to which sustainability competencies develop.

The WSA represents a transformative model for embedding sustainability into the core of educational institutions. Rather than treating sustainability as an isolated subject in the curriculum or a one-off project, the WSA integrates sustainability into all dimensions of school life: leadership and governance, teaching and learning, community engagement, school operations, and school culture and values.

This approach responds to a growing consensus in education and sustainability research that addressing complex, systemic challenges—such as climate change, biodiversity loss, and social inequality—requires educational responses that are themselves systemic and integrated. The WSA aligns with the goals of frameworks like SDG 4.7 and GreenComp by fostering sustainability competences through participatory learning environments, cross-curricular integration, and schoolwide coherence (Mathie & Wals, 2022).

Importantly, the WSA encourages schools to become living laboratories of sustainability. This means engaging not just in curricular reform but also in organizational and relational transformation—reconfiguring how schools operate, how decisions are made, and how the school interacts with its wider community. Schools applying a WSA "walk the talk" by aligning their physical infrastructure (for example, energy use, waste, food) and their social infrastructure with the sustainability values they aim to promote (Verhelst et al., 2024).

The power of WSA lies in its capacity to develop competencies such as systems thinking, complexity handling, ethical reasoning and empathy, care, reflexivity, and hope. It embraces the notion that students must not only learn about sustainability but learn to live sustainably—within ecological boundaries and guided by ethical and democratic values.

Ultimately, the WSA recognizes that "education-as-usual" is insufficient in the face of planetary challenges. It calls for a reorientation of schooling itself—toward being a driver of social transformation and ecological regeneration.

Understanding how students develop these competences requires robust, comparable data on both what they know and their attitudes regarding the environment. TIMSS 2023 introduces a dedicated Environmental Attitudes and Behaviors Framework (Reynolds & Komakhidze, 2022), alongside its traditional science assessment, enabling the measurement of both environmental knowledge (through cognitive science items) and environmental attitudes and behaviors through contextual questionnaires.

To operationalize the WSA in real educational settings, it is essential to identify the organizational conditions that enable its implementation. Drawing on Verhelst et al. (2020), the ESD-effective school framework identifies eight critical school-level characteristics (Fig. 6.1). At the foundational level, the interplay between school resources and sustainable leadership creates the conditions necessary for six core organizational dimensions: pluralistic communication, supportive relationships within and beyond the school, collective efficacy, adaptability, democratic decision-making, and a shared vision. Together, these factors form an enabling school culture for effective sustainability education.

Appendix 6.1 includes a table mapping these characteristics to relevant scales from the TIMSS 2023 school context questionnaire, providing a way to explore how

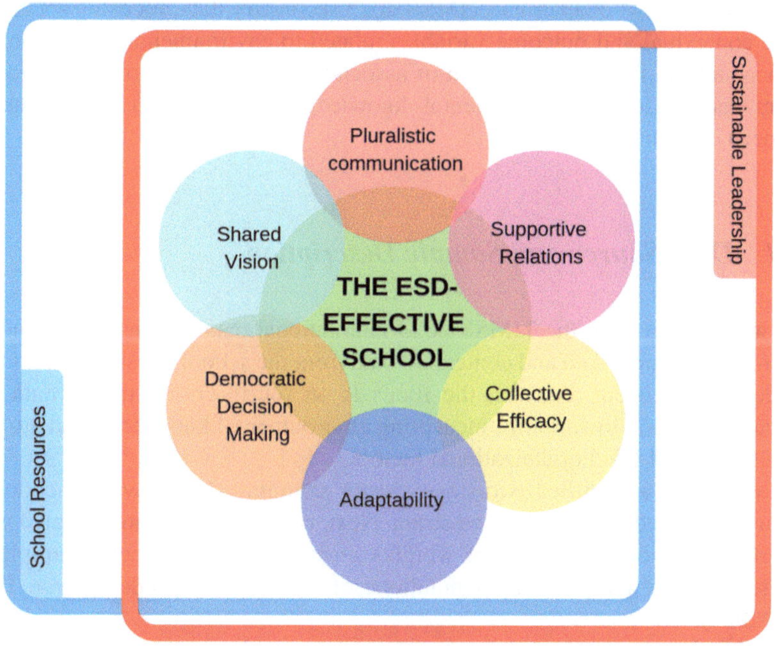

Fig. 6.1 ESD-effective school framework. Source: Reproduction of *Fig. 1: a representation of the framework for an ESD-effective school* from Verhelst et al. (2020, p. 13)

schools across education systems foster, or fail to foster, an environment conducive to sustainability competence development. While several of the characteristics are covered by the TIMSS 2023 school questionnaire, some characteristics of the framework are not reflected.

These characteristics are interconnected, and they shape schools' organizational culture. Establishing such an enabling context may not only be crucial for the effective implementation of a WSA but be a necessary condition to achieve impactful sustainability education.

However, school organization and resources can also be affected by other external factors such as the location and socioeconomic context. This discussion has gained increasing relevance due to the ongoing climate crisis (Chap. 7). Previous studies have shown that urban and rural settings may differ in terms of students' environmental knowledge and attitudes. For instance, research suggests that these differences are often shaped by varying degrees of exposure to environmental issues, disparities in resource availability, and the contextual relevance of sustainability education across different localities (Echazarra & Radinger, 2019; Strello et al., 2025).

6.3 Methodology

The analyses in this chapter use TIMSS 2023 secondary data, focusing specifically on the contextual and outcome variables related to environmental education. This approach allows for the exploration of associations between school organizational factors and students' environmental knowledge and attitudes across multiple European education systems.

6.3.1 Data Source and Sample Description

Data were drawn from the TIMSS 2023 international database, using fourth grade student achievement data and contextual data from the school questionnaire. For the broader cross-system analysis, the focus is on fourth-grade student data, with particular attention to variables measuring environmental knowledge and attitudes, as well as school-level organizational factors.

TIMSS uses a stratified two-stage cluster sampling design where schools are randomly selected within specified strata. Next, random classes are sampled within the selected schools. The use of weights ensures that the selected samples are representative of the national populations of fourth-grade students (von Davier et al., 2024).

TIMSS data are collected from fourth-grade students, their teachers, school principals, and parents. These datasets allow for an analysis of how different learning environments shape environmental competences and enable cross-national comparisons, particularly within the European context.

The cross-national analysis in this chapter includes fourth-grade data from 28 education systems within the European Education Area and the Western Balkans, specifically: Albania, Belgium (Flemish), Belgium (French), Bosnia and Herzegovina, Bulgaria, Cyprus, the Czech Republic, Denmark, Finland, France, Germany, Hungary, Ireland, Italy, Kosovo, Latvia, Lithuania, Montenegro, the Netherlands, the Republic of North Macedonia, Poland, Portugal, Romania, Serbia, the Slovak Republic, Slovenia, Spain, and Sweden. This chapter uses the data from the school questionnaire and student context questionnaire, and students' achievement data (Mullis et al., 2021; Reynolds & Komakhidze, 2022).

6.3.2 Variables and Operationalization

The study included the following variables: students' environmental knowledge and attitudes assessed using the Environmental Awareness Achievement scale and Students Value Environmental Preservation scale from the student achievement test and the student context questionnaire, respectively (von Davier et al., 2024), and contextual factors drawn from the TIMSS 2023 school questionnaire (Table 6.1). These variables were operationalized in alignment with the ESD-effective school framework (Annex 6.1).

For clearer comparative analysis between urban and rural areas and to increase statistical power within subgroups, schools have been grouped into two main categories: schools in more urbanized areas and schools in more rural areas.

Table 6.1 Variables included in the study drawn from TIMSS 2023 school and student questionnaires

Variable type	Category	Variable(s)	Description/source
Independent variables	Background/ contextual factors	ACBG05A, ACBG05B	People in area and immediate school surroundings
	Resources	ACBGSRS, ACBG08, ACBG11C	Instruction affected by science resource shortage and resource availability including science labs, equipment and facilities
	School emphasis on sustainability	ACBG13A–H	Shared vision, schoolwide approach, teach to appreciate nature and act, community engagement and teacher support
Criterion variables	Environmental outcomes	ASSENV0	Environmental knowledge (scale from cognitive test items)
		ASGDVEP	Environmental attitudes (scale on valuing environmental preservation)

6.3.3 *Analytical Procedures*

The main data analyses were conducted using the IDB (International Database) Analyzer, developed by the IEA for international large-scale assessments (IEA, 2021). This ensures proper use of the plausible values and weights which are common in TIMSS data. To address the first research question, descriptive statistics were used to explore cross-national differences in students' environmental knowledge, attitudes, and school-level emphasis on sustainability. In addition, the intraclass correlation (ICC) was calculated using the *WeMix* package in R (Bailey et al., 2025; Huang, 2024). To address the second and third research questions, Pearson correlations examined associations between school organizational factors and student outcomes, with attention to both statistical significance and practical relevance (Cohen, 1988). All analyses followed TIMSS sampling and weighting procedures to ensure valid inferences (Johansone, 2024). Given the exploratory nature of the research questions, the analyses were used to identify patterns and relationships, rather than to test specific hypotheses.

6.4 Results

6.4.1 *Environmental Knowledge and Attitudes*

This section presents cross-system differences in students' environmental knowledge and attitudes and explores how school-level emphasis on sustainability varies among participating education systems.

Environmental knowledge (ASSENV0) reflects students' science performance related to environmental topics,[1] whereas environmental attitudes (ASDGVEP) were measured using the Students Value Environmental Preservation scale.[2]

Denmark, Finland, Ireland, and Poland stand out for high levels of environmental knowledge, while Bosnia and Herzegovina, Kosovo, and the Republic of North Macedonia score the lowest (Fig. 6.2a). Interestingly, the pattern is reversed for environmental attitudes: Albania, Bulgaria, and Romania show the strongest pro-environmental attitudes, whereas Denmark, Finland and Sweden—despite their high knowledge levels—rank lower in this domain (Fig. 6.2b).

This contrast in score values suggests that knowledge and attitudes do not always align. For example, Nordic education systems perform well in knowledge but less so in attitudes, while several Eastern European education systems show the opposite. The average environmental attitudes score is 10.14, with a relatively narrow range (9.0–11.0), and the average knowledge score is 504, indicating generally positive knowledge and attitudes across education systems.

The comparative analysis of school emphasis on environmental education across European education systems (Fig. 6.3 and Fig. 6.4) reveals a consistent regional

[1] Scale with a set mean of 500 and a standard deviation of 100.

[2] Scale with a mean of 00 and a standard deviation of 00 within the subsample of education systems.

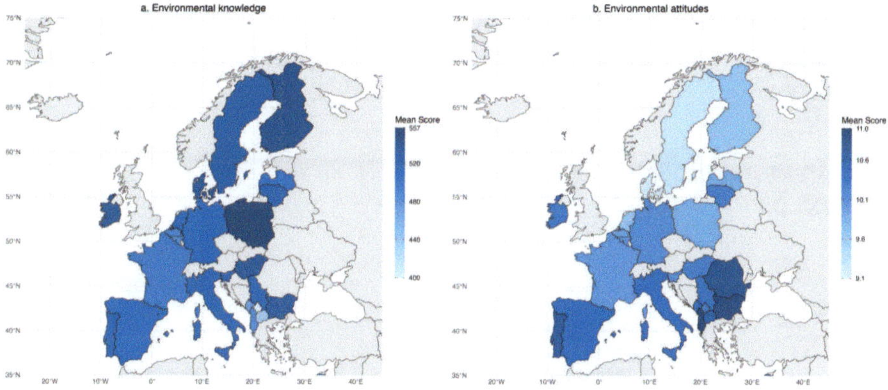

Fig. 6.2 Environmental knowledge (ASSENV0) and attitudes (ASDGVEP) scores by education system

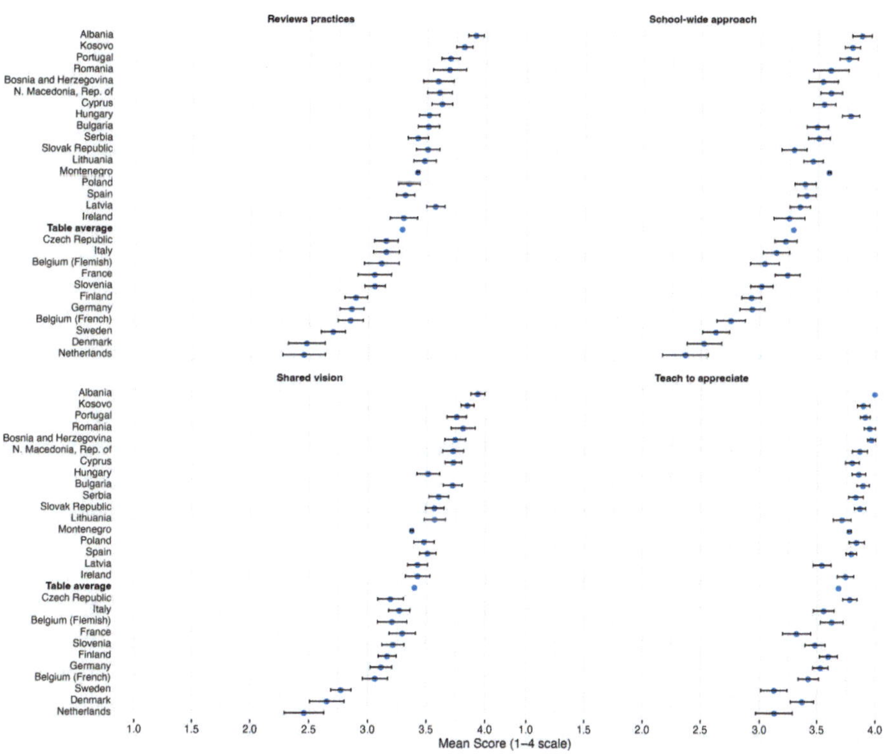

Fig. 6.3 School emphasis on sustainability: ACBG13A–ACBG13D

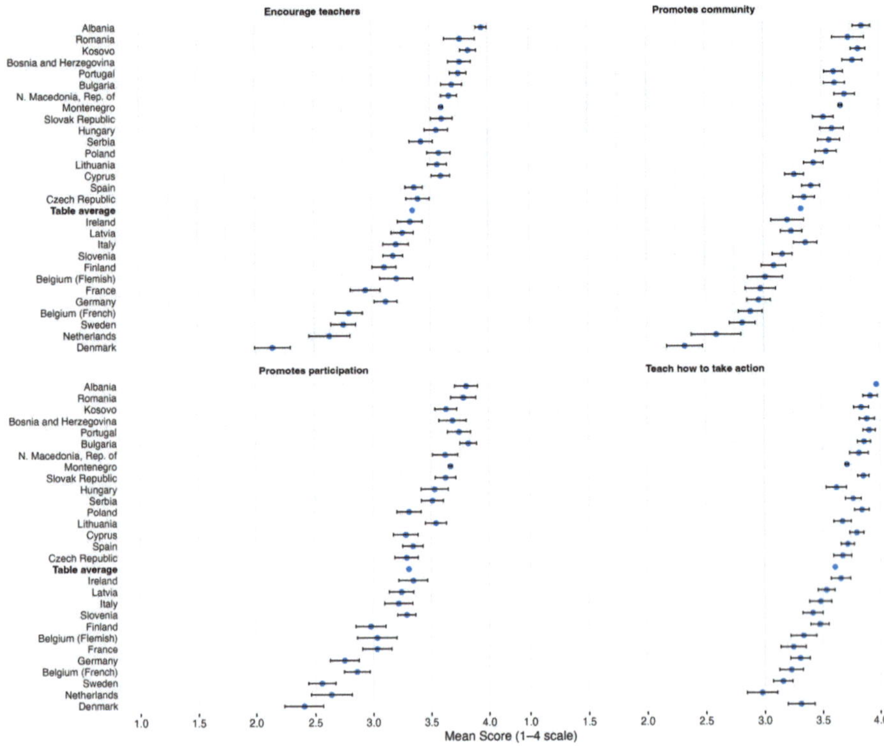

Fig. 6.4 School emphasis on sustainability: ACBG13E–ACBG13H

divide. Eastern and Southeastern European education systems—such as Albania, Kosovo, and Romania—demonstrate significantly higher levels of emphasis across nearly all examined dimensions, including strategic practices (for example, reviewing environmental initiatives, fostering a shared vision), community engagement, and student participation. In contrast, Western and Northern European education systems—such as Denmark, Germany, the Netherlands, and Sweden—tend to rank lower, especially in structural and institutional aspects like encouraging teachers or implementing a schoolwide approach.

In addition, the ICC for the different education systems for both environmental knowledge (ASSENV0) and environmental attitudes (ASDGVEP) (Table 6.2) shows that the variance in these variables by school level differs greatly across education systems. In general, the school level tends to explain less variance in environmental knowledge than it does in attitudes. Despite the differences between education systems, the school level seems to play an important part in explaining differences in students' outcomes related to environmentalism, especially regarding environmental knowledge.

Table 6.2 ICC for environmental knowledge (ASSENV0) and attitudes (ASDGVEP) for the different education systems

Education system	ICC value	
	ASSENV0	ASDGVEP
Bulgaria	0.589	0.185
Slovak Republic	0.405	0.104
Hungary	0.353	0.056
Romania	0.299	0.058
Germany	0.293	0.64
Albania	0.280	0.64
N. Macedonia, Rep. of	0.273	0.081
Belgium (Flemish)	0.239	0.054
Kosovo	0.239	0.050
Lithuania	0.219	0.042
Bosnia and Herzegovina	0.205	0.032
Sweden	0.195	0.038
Cyprus	0.188	0.037
Spain	0.184	0.064
Belgium (French)	0.182	0.047
Serbia	0.180	0.066
Ireland	0.175	0.026
Czech Republic	0.173	0.032
Montenegro	0.171	0.037
Italy	0.169	0.044
Portugal	0.169	0.037
France	0.163	0.041
Finland	0.149	0.044
Netherlands	0.149	0.028
Latvia	0.144	0.054
Denmark	0.132	0.054
Poland	0.095	0.033
Slovenia	0.055	0.035

While variables like "teach to appreciate" and "teach how to take action" receive generally higher ratings across the board (Figs. 6.3 and 6.4), even here the East–West gap persists. The Belgium (Flemish) case specifically illustrates a pattern of moderate-to-low emphasis across both strategic and participatory dimensions, suggesting that while environmental values may be acknowledged, they are not strongly embedded in schoolwide practices or supported through staff empowerment. These findings highlight potential areas for policy attention, particularly in education systems where structural support and participatory strategies for environmental education remain underdeveloped.

6.4.2 Relation Between the School Area and Environmental Knowledge and Attitudes

This section explores how the geographic location of schools—urban versus rural—affects students' environmental knowledge and attitudes. Building on the prior analysis of cross-system differences in environmental knowledge, this section focuses specifically on the role of school setting. The original dataset categorized schools into five types (Rural, Small Town, Mid-City, Suburban, and Urban), which were recodified into two broader groups: rural (Rural and Small Town) and urban (Mid-City, Suburban, and Urban). This simplification allows for a clearer comparison of students' environmental literacy across these two contexts. To quantify these differences, the average environmental knowledge and attitude scores were calculated for each education system and school area type. Confidence intervals were also computed to assess the reliability of the observed gaps.

The urban–rural gaps in students' environmental knowledge and attitudes across European education systems were examined (Fig. 6.5). Environmental knowledge is measured on a science performance scale using item response theory, where the international mean is typically centered around 500, while environmental attitudes is measured on a more compressed scale derived from student responses to the Students Value Environmental Preservation scale.

Regarding urban–rural gaps in environmental knowledge (Fig 6.5a), education systems like Bulgaria, Hungary, and the Slovak Republic show pronounced gaps favoring urban areas, with urban students scoring higher than their rural peers. On the other hand, France, Ireland, and Portugal display smaller or even slightly reversed gaps, with rural students scoring slightly better. Additionally, education systems including the Czech Republic, Finland, the Republic of North Macedonia, and Slovenia display only modest differences in environmental knowledge between urban and rural students. This indicates a more equitable educational context, where

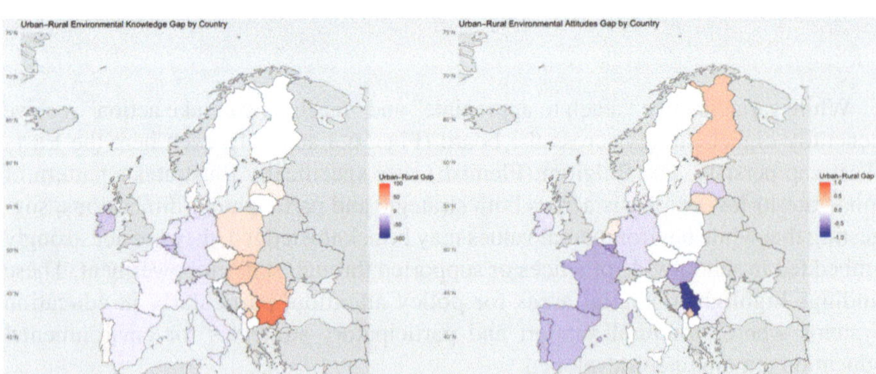

Fig. 6.5 Urban-rural gaps in environmental knowledge and attitudes across European education systems. Note: Red tones indicate education systems where urban students perform better, and blue tones indicate education systems where rural students perform better

school location does not substantially influence students' environmental knowledge. Belgium presents a nuanced picture, with differences between its Flemish and French communities. In terms of environmental knowledge, both communities show relatively small urban–rural gaps. The Flemish community displays a slight urban advantage, while the French community appears nearly balanced.

Regarding urban–rural gaps in environmental attitudes (Fig. 6.5b), a more complex and less uniform picture is revealed. While education systems like Finland and Lithuania exhibit a moderate urban advantage in environmental attitudes, with urban students expressing more concern about environmental preservation, the trend is not consistent across the continent. Notably, Serbia demonstrates a strong rural advantage, with rural students reporting significantly more pro-environmental attitudes than their urban peers. This pattern is also visible, although less intensely, in education systems like France, Portugal, and Spain, suggesting that in certain contexts, rural students may develop a deeper connection to nature and environmental values. In Belgium, the French community shows a more noticeable rural advantage in environmental attitudes while the Flemish community shows minimal differences in attitudes between urban and rural areas, indicating a more balanced attitudinal profile.

Overall, the influence of school location on environmental knowledge and attitudes is neither uniform nor straightforward across Europe. In some education systems, particularly Eastern Europe, urban students appear to benefit more in terms of environmental knowledge, potentially due to greater access to educational resources or extracurricular opportunities. In contrast, the distribution of environmental attitudes is more mixed, with some rural areas—such as in Serbia and parts of Western Europe—showing higher environmental attitudes.

6.4.3 School Emphasis on Sustainability and the Relationship with Knowledge and Attitudes

Various school emphasis dimensions, such as shared vision (SSV), whole school approach (WSA), and teacher encouragement (ENT), are moderately to strongly correlated with each other (ranging from 0.44 to 0.71) (Fig. 6.6). This suggests that schools that place a significant emphasis on one aspect of environmental education tend to prioritize others as well, implying a coherent or integrated approach to promoting environmental awareness and education within the institution.

However, when examining the relationship between these school emphasis variables and students' environmental attitudes, the findings show very weak or no significant correlations (ranging from −0.01 to −0.02) (Fig. 6.6). This weak association implies that a school's strategic focus on environmental education does not have a strong impact on how students perceive or think about environmental issues, at least not as measured by the environmental attitudes (ASDGVEP) tool. This could suggest that school-level organizational factors play a limited role in shaping

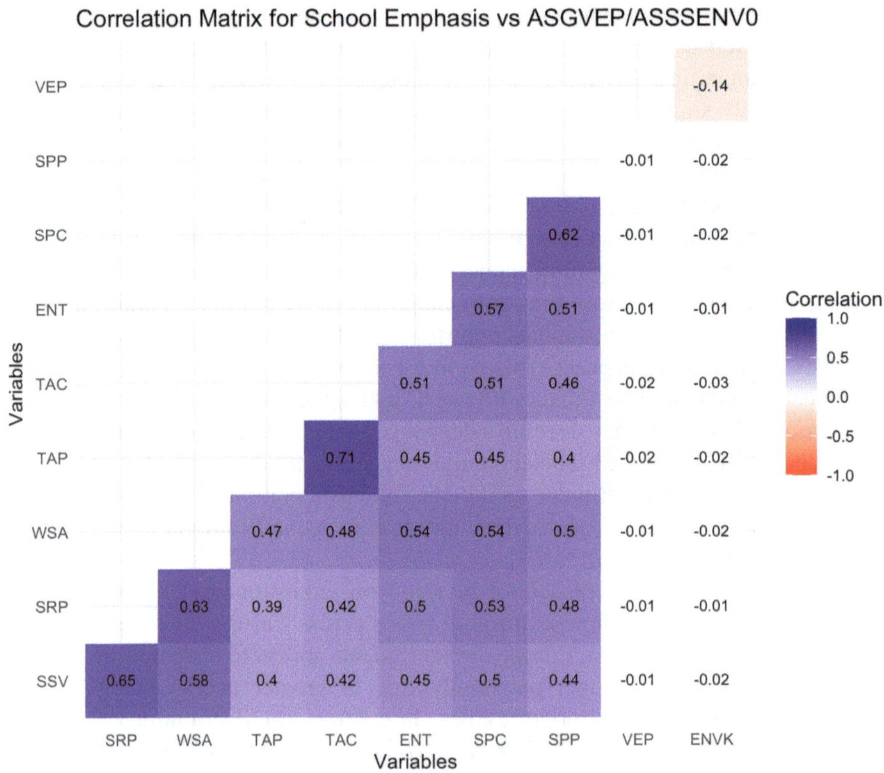

Fig. 6.6 Correlation matrix between school emphasis on sustainability and knowledge/attitudes. Notes: *ENVK* students' environmental knowledge, *VEP* students' environmental attitudes, *SPP* school promotes participation, *SPC* school promotes community involvement, *ENT* teacher encouragement, *TAC* teach how to take action, *TAP* teach to appreciate environment, *WSA* whole school approach, *SRP* school reviews practices, *SVV* school shared vision

students' environmental attitudes, with personal, familial, or societal influences potentially having a stronger effect.

An interesting observation is the small negative correlation between environmental knowledge (ENVK) and environmental attitudes (VEP), with a value of −0.14 (Fig. 6.6). While the relationship is not strong, it suggests that students with more environmental knowledge may have slightly less idealistic attitudes towards environmental issues. This could indicate that a higher level of knowledge does not necessarily lead to more optimistic or positive views, or it may suggest that idealistic attitudes do not always correlate with accurate environmental understanding. Alternatively, this inverse relationship might reflect cultural differences in how knowledge and attitudes towards environmental issues are formed and expressed.

6.4.4 School Resources and the Relationship with Environmental Knowledge and Attitudes

This section examines how students' perceptions of school resources relate to their environmental knowledge and attitudes across education systems. Analysis of the correlation between student knowledge and perceived school resources across education systems, reveals a generally positive but consistently low relationship (Fig. 6.7). In all cases, the correlations remain below 0.15, indicating only weak associations.

These findings suggest that while there is a modest tendency for students in better-resourced schools to demonstrate higher knowledge levels, the strength of this relationship is minimal across all education systems. A correlation of 0.10, even if statistically significant, reflects only a very weak association in practical terms. Compared to the typically weaker correlations between attitudes and knowledge, the slightly more consistent significance here points to a somewhat stronger link between resources and cognitive outcomes.

Nonetheless, the small magnitude of all correlations cautions against attributing student learning outcomes to school resources alone.

When examining the relationship between students' perceptions of school resources and their attitudes across various education systems (Fig. 6.8), while Bulgaria, France, Hungary, Ireland, and the Republic of North Macedonia report positive and statistically significant correlations, these correlations are notably weak (below 0.10). Most education systems exhibit non-significant correlations, indicating no discernible or consistent linear relationship between perceived school

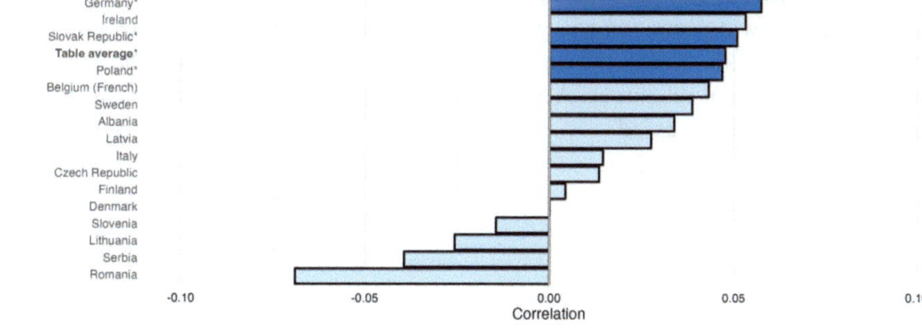

Fig. 6.7 Correlation plot between school resources and knowledge by education system. Note: * Indicates $p < 0.05$

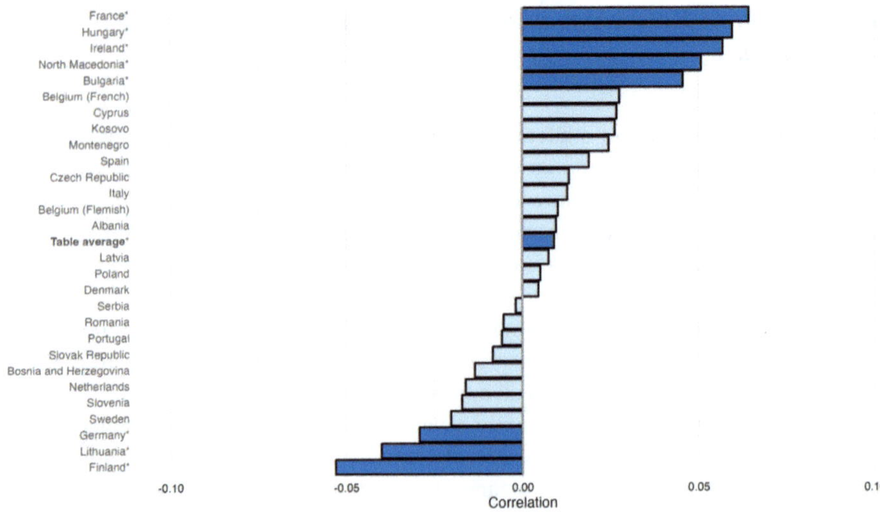

Fig. 6.8 Correlations between school resources and attitudes by education system. Note: * Indicates $p < 0.05$

resources and student attitudes. This points to the influence of other factors—such as teaching quality, peer relationships, or curriculum content—that may play a stronger role in shaping students' environmental attitudes.

6.5 Discussion

This chapter set out to explore how school-level variables relate to student outcomes in environmental knowledge and attitudes across European Education Area and Western Balkans education systems. The exploratory nature of this analysis using TIMSS 2023 data allowed identification of patterns and relationships between school-level variables and student outcomes, as a first step in investigating the nature of the role of the schools within the TIMSS environmental awareness framework. The following sections will go over the main findings of this study, related to the urban–rural differences, the limiter role of the school-level organizational variables and the school resources. Afterwards, the implications for future assessment and policy will be discussed.

6.5.1 Key Findings

A key finding of this study is the cross-national variation in both environmental knowledge and attitudes among fourth-grade students across the European Education Area and Western Balkans. Intriguingly, a weak negative correlation was observed between environmental knowledge and environmental attitudes. This counterintuitive result aligns with previous research of TIMSS and International Civic and Citizenship Education Study (ICCS) data (Isac et al., 2025; Kollmuss & Agyeman, 2002). In addition, the ICC indicates that the role of the school organization in explaining variance in the outcome variables related to environmentalism differs across education systems. Several factors might contribute to this disconnect. One possible explanation lies in the nature of knowledge assessed by TIMSS 2023, which focusses primarily on cognitive science items. These might not fully capture environmental attitudes. Furthermore, cultural contexts across the diverse nations studied may play a significant role in shaping how environmental attitudes are formed, valued, and expressed, independent of formal knowledge. Education efforts around sustainability should aim to cultivate not only robust environmental literacy but also engagement and action-oriented learning contextualized to the local context of the student, as envisioned by frameworks like GreenComp.

6.5.1.1 Urban–Rural Differences in Environmental Outcomes

The analysis of urban versus rural school settings reveals a complex and non-uniform landscape across Europe regarding environmental knowledge and attitudes. In several Eastern European education systems, urban students demonstrated higher levels of environmental knowledge. This could be attributed to potentially greater access to diverse educational resources, specialized science instruction, or extracurricular environmental programs often concentrated in urban areas. In contrast, the pattern for environmental attitudes was more varied, with some rural areas, such as in Serbia, exhibiting stronger environmental attitudes. This may stem from rural students having more direct and frequent interactions with the natural environment or other factors which may need to be investigated with more detail.

The divergence between knowledge and attitudes in these different settings is particularly interesting and further correlations with other factors could be studied in the future. Although the gap between urban and rural setting is high in some education systems (eastern education systems show a relative larger gap in environmental knowledge) and small in others (France and Spain have a relative smaller gap in environmental attitudes). Nevertheless, this gap does not have a direct relation to the mean values in these education systems, indicating that other contextual factors may be of importance.

6.5.1.2 Limited Role of School-Level Organizational Variables

A second interesting finding of this chapter is the limited explanatory power of school-level organizational factors when it comes to students' environmental knowledge and attitudes. Correlations between the emphasis placed by schools on environmental issues and the outcomes in student knowledge or attitudes is very low. While schools that prioritize certain sustainability aspects (for example, having a shared vision or encouraging teachers) tend to do so across other dimensions as well—suggesting internal coherence in their organizational culture—this internal alignment does not seem to translate into significantly higher student outcomes. This disconnect raises questions about the mechanisms through which school-level strategies and characteristics influence student learning or affective development in the field of sustainability.

6.5.1.3 School Resources and Environmental Outcomes

Third, the role of school resources, measured through indicators such as access to science equipment or materials, was found to have a weak but consistently positive correlation with environmental outcomes. Across education systems, this relationship was slightly stronger in the case of environmental knowledge than attitudes but never exceeded modest levels of correlation. This suggests that while having resources may facilitate more effective teaching in science and environmental domains, it is far from a determining factor. Attitudes may be more strongly influenced by social, cultural, and familial variables beyond the reach of school provisioning.

6.5.2 Implications for Future Assessment and Policy

The most probable explanation for these unexpected observations is that TIMSS 2023 is an academic achievement-focused assessment; it primarily captures cognitive outcomes, particularly those aligned with scientific knowledge and reasoning. This narrow cognitive orientation limits the framework's ability to fully represent the multidimensional competences outlined in frameworks such as the WSA. The framework emphasizes not just knowledge, but values, attitudes, agency, and the capacity for systemic thinking, competencies that are often developed through schoolwide processes and participatory practices not directly measured by TIMSS (Nusche et al., 2024).

While this chapter's exploratory approach enabled a broad examination of the relationships between school-level variables and student outcomes, it also inherently limits the strength of the conclusions that can be drawn. Future research could build on this foundation by exploring how the TIMSS 2023 variables and the environmental awareness framework align with established models in environmental and sustainability education, such as the ESD-effective school framework or the WSA. Additionally, the international comparisons presented in this chapter warrant further investigation, particularly through the assessment of measurement invariance across the constructs within the TIMSS environmental awareness framework.

The WSA, which promotes a holistic and process-oriented vision of sustainability education, emphasize the transformation of school culture, leadership, community engagement, and student empowerment. Its core logic is not primarily about what students know, but about how they live and act within learning environments that model sustainability in everyday practice. Although the ESD-effective school framework tries to use the items from the TIMSS 2023 assessment to understand the relationship between school organization and sustainability knowledge and attitudes, the nature of the assessment or the selection of items does not show important correlations.

Based on this misalignment, and taking into account previous studies that did find positive correlations between school-level variables and student outcomes in sustainability education (Boeve-de Pauw & Van Petegem, 2018; Verhelst et al., 2022), it is recommend that future assessments should address not only individual cognitive outcomes, but also the institutional and pedagogical processes that underpin sustainability learning. Integrating tools that capture the transformational dimensions of the WSA, such as leadership for sustainability, collaborative decision-making, and values-driven community engagement, would offer a more comprehensive picture of how environmental competences are fostered in schools.

This chapter underscores the need to conceptualize schools as complex systems whose potential to foster environmental competence extends beyond curricular content or resource allocation. As shown in the analysis, neither school emphasis on sustainability nor the availability of resources alone can account for variations in student attitudes or knowledge. What appears to matter more are the organizational dynamics and cultural patterns within schools—shared vision, collective efficacy, adaptability, and supportive relationships—that shape how sustainability is embedded and lived in everyday school life. A more nuanced understanding and measurement of these processes is crucial for education systems seeking to prepare learners for the environmental challenges of the twenty-first century.

Appendix (Table 6.3)

Table 6.3 Mapping of ESD-effective school characteristics to TIMSS 2023 items

Characteristic of an ESD-effective school	Related scales from the TIMSS 2023 school context questionnaire
Sustainable leadership	N/A
School resources	Scale "instruction affected by science resource shortage"
Pluralistic communication	N/A
Supportive relations	• This school encourages teachers to learn about teaching environmental sustainability • This school promotes environmental sustainability within the community • This school promotes participation in networks or projects related to environmental sustainability
Collective efficacy	• This school teaches students to appreciate the natural environment • This school teaches students how to take action to protect the natural environment
Adaptability	This school regularly reviews its practices to make them more environmentally sustainable
Democratic decision-making	N/A
Shared vision	• This school has a shared vision about its role in supporting environmental sustainability • There is a schoolwide approach to teaching environmental sustainability to students

References

Bailey, P., Webb, B., Kelley, C., Nguyen, T., & Huo, H. (2025). *WeMix: Weighted mixed-effects models using multilevel pseudo maximum likelihood estimation.* https://american-institutes-for-research.github.io/WeMix/

Ballegeer, A. M., Ferrari, E., Corrochano, D., & Ruiz, C. (2024). The role of knowledge, attitude, and emotions in Spanish pre-service teachers' willingness to engage in climate change education. *European Journal of Teacher Education*, 1–19. https://doi.org/10.1080/02619768.2024.2370885

Bianchi, G., Pisiotis, U., & Cabrera Giraldez, C. (2022). *GreenComp: The European sustainability competence framework*. European Commission: Joint Research Centre. Publications Office of the European Union. https://doi.org/10.2760/13286

Boeve-de Pauw, J., Gericke, N., Olsson, D., & Berglund, T. (2015). The effectiveness of education for sustainable development. *Sustainability, 7*(11), 15693–15717. https://doi.org/10.3390/su71115693

Boeve-de Pauw, J., & Van Petegem, P. (2018). Eco-school evaluation beyond labels: The impact of environmental policy, didactics and nature at school on student outcomes. *Environmental Education Research, 24*(9), 1250–1267. https://doi.org/10.1080/13504622.2017.1307327

Borgonovi, F., Brussino, O., Seitz, H., Bertoletti, A., Biagi, F., Bitat, A., Karpinski, Z., & Montanari, M. (2022). The environmental sustainability competence toolbox: From leaving a better planet for our children to leaving better children for our planet. *OECD Social, Employment and Migration Working Papers*, No. 275. OECD Publishing. https://doi.org/10.1787/27991ec0-en.

Cohen, J. (1988). *Statistical power analysis for the behavioral sciences*. Routledge. https://doi.org/10.4324/9780203771587

Echazarra, A., & Radinger, T. (2019). Learning in rural schools. *OECD Education Working Papers*, No. 196. OECD Publishing. https://doi.org/10.1787/8b1a5cb9-en.

European Commission. (2019). The European Green Deal. COM(2019) 640 final. https://eur-lex.europa.eu/legal-content/EN/TXT/?uri=celex:52019DC0640

Huang, F. L. (2024). Using plausible values when fitting multilevel models with large-scale assessment data using R. *Large-scale Assessment in Education, 12*(7). https://doi.org/10.1186/s40536-024-00192-0

IEA. (2021). *Data and tools. Tools.* https://www.iea.nl/data-tools/tools

Intergovernmental Panel on Climate Change. (2023). *Climate change 2023: Synthesis report.* Contribution of Working Groups I, II and III to the Sixth Assessment Report of the Intergovernmental Panel on Climate Change [Core Writing Team, H. Lee & J. Romero (Eds.)]. https://doi.org/10.59327/IPCC/AR6-9789291691647

Isac, M. M., Sandoval-Hernández, A., & Sass, W. (Eds.). (2025). *Knowledge and willingness to act pro-environmentally. Perspectives from IEA TIMSS 2019 and ICCS 2016 data* (IEA research for education) (Vol. 16). Springer. https://doi.org/10.1007/978-3-031-76033-4

Johansone, I. (2024). TIMSS survey operations procedures. In M. von Davier, B. Fishbein, & A. Kennedy (Eds.), *TIMSS 2023 technical report (methods and procedures)* (pp. 4.1–4.19). TIMSS & PIRLS International Study Center, Boston College. https://doi.org/10.6017/lse.tpisc.timss.rs2399

Klieme, E., & Nilsen, T. (2022). Teaching quality and student outcomes in TIMSS and PISA. In T. Nilsen, A. Stancel-Piątak, & J. E. Gustafsson (Eds.), *International handbook of comparative large-scale studies in education* (Springer international handbooks of education). Springer. https://doi.org/10.1007/978-3-030-88178-8_37

Kollmuss, A., & Agyeman, J. (2002). Mind the gap: Why do people act environmentally and what are the barriers to pro-environmental behavior? *Environmental Education Research, 8*(3), 239–260. https://doi.org/10.1080/13504620220145401

List, M. K., Schmidt, F. T. C., Mundt, D., & Föste-Eggers, D. (2020). Still green at fifteen? Investigating environmental awareness of the PISA 2015 population: Cross-national differences and correlates. *Sustainability, 12*(7), 2985. https://doi.org/10.3390/su12072985

Mathie, R. G., & Wals, A. E. J. (2022). *Whole school approaches to sustainability: Exemplary practices from around the world.* Wageningen University. https://doi.org/10.18174/566782

Mogren, A., Gericke, N., & Scherp, H.-Å. (2019). Whole school approaches to education for sustainable development: A model that links to school improvement. *Environmental Education Research, 25*(4), 508–531. https://doi.org/10.1080/13504622.2018.1455074

Mullis, I. V. S., Martin, M. O., & von Davier, M. (Eds.). (2021). *TIMSS 2023 assessment frameworks.* TIMSS & PIRLS International Study Center, Boston College. https://timssandpirls.bc.edu/timss2023/frameworks/index.html

Nusche, D., Rabella, M. F., & Lauterbach, S. (2024). Rethinking education in the context of climate change: Leverage points for transformative change. *OECD Education Working Papers*, No. 307. OECD Publishing. https://doi.org/10.1787/f14c8a81-en

Reynolds, K. A., & Komakhidze, M. (2022). *TIMSS 2023 environmental attitudes and behaviors framework.* TIMSS & PIRLS International Study Center, Boston College. https://timssandpirls.bc.edu/timss2023/frameworks/pdf/T23-Environmental-Attitudes-and-Behaviors-Framework.pdf

Sass, W., Boeve-de Pauw, J., Olsson, D., Gericke, N., De Maeyer, S., & Van Petegem, P. (2020). Redefining action competence: The case of sustainable development. *The Journal of Environmental Education, 51*(4), 292–305. https://doi.org/10.1080/00958964.2020.1765132

Scott, W. (2013). Developing the sustainable school: Thinking the issues through. *The Curriculum Journal, 24*(2), 181–205. https://doi.org/10.1080/09585176.2013.781375

Strello, A., Strietholt, R., Liaw, Y. L., Baghaei, P., & Meinck, S. (2025). Inequalities on environmental knowledge. In M. M. Isac, A. Sandoval-Hernández, & W. Sass (Eds.), *Knowledge and willingness to act pro-environmentally* (IEA research for education) (Vol. 16, pp. 57–76). Springer. https://doi.org/10.1007/978-3-031-76033-4_5

UNESCO. (2017). *Education for sustainable development goals: Learning objectives*. UNESCO Publishing. https://doi.org/10.54675/CGBA9153

UNESCO. (2022). *Berlin Declaration on Education for Sustainable Development. Learn for our planet: Act for sustainability*. UNESCO Publishing. https://unesdoc.unesco.org/ark:/48223/pf0000381228

UNESCO. (2024). *Education and climate change: Learning to act for people and planet*. UNESCO Publishing. https://doi.org/10.54676/GVXA4765

United Nations. (2015). *Transforming our world: The 2030 Agenda for Sustainable Development* (A/RES/70/1). https://sdgs.un.org/2030agenda

Verhelst, D., Mogren, A., Boeve-de Pauw, J., & Van Petegem, P. (2024). Getting to a whole school approach: Lessons from school effectiveness and school improvement in ESD research. In A. E. Wals, B. Bjønness, A. Sinnes, & I. Eikeland (Eds.), *Whole school approaches to sustainability* (Sustainable development goals series) (pp. 71–86). Springer. https://doi.org/10.1007/978-3-031-56172-6_5

Verhelst, D., Vanhoof, J., Boeve-de Pauw, J., & Van Petegem, P. (2020). Building a conceptual framework for an ESD-effective school organization. *The Journal of Environmental Education, 51*(6), 400–415. https://doi.org/10.1080/00958964.2020.1797615

Verhelst, D., Vanhoof, J., de Maeyer, S., Sass, W., & Van Petegem, P. (2022). Enabling effective education for sustainable development: Investigating the connection between the school organization and students' action competence. *The Journal of Environmental Education, 53*(4), 171–185. https://doi.org/10.1080/00958964.2022.2072797

Verhelst, D., Vanhoof, J., & Van Petegem, P. (2023). School effectiveness for education for sustainable development (ESD): What characterizes an ESD-effective school organization? *Educational Management Administration and Leadership, 51*(2), 502. https://doi.org/10.1177/1741143220985196

von Davier, M., Fishbein, B., & Kennedy, A. (Eds.). (2024). *TIMSS 2023 technical report (methods and procedures)*. TIMSS & PIRLS International Study Center, Boston College. https://timss2023.org/methods

Wals, A. E. J., Bjønness, B., Sinnes, A., & Eikeland, I. (2024). *Whole school approaches to sustainability. Education renewal in times of distress*. Springer. https://doi.org/10.1007/978-3-031-56172-6

White, P., Ardoin, N. M., Eames, C., & Monroe, M. C. (2023). Agency in the Anthropocene: Supporting document to the PISA 2025 science framework. *OECD Education Working Papers*, No. 297. OECD Publishing. https://doi.org/10.1787/8d3b6cfa-en.

Inés García-Bohórquez is a PhD candidate at the University of Salamanca, affiliated with the EMC3 research group in the Department of Mathematics and Science Education. Her research focuses on climate change education through theoretical and statistical modeling using ILSA data. She holds a degree in Primary Education and a master's in ICT in Education.

Dries Verhelst is a post-doc researcher in the Faculty of Social Sciences at the University of Antwerp. His research includes school policy and school organizational management, educational effectiveness, and school effectiveness, with a focus on sustainability education. He is a member of the Edubron research group.

Fernando Martínez-Abad is an Associate Professor in Research Methods in Education at the University of Salamanca and Director of the Master's Degree in Evaluation and Research in Learning Organizations and Contexts.

His main research interests include factors related to academic achievement and the analysis of large-scale international educational assessments, with a focus on school effectiveness and educational equity.

Camilo Ruiz is a Professor in the Department of Science Education at the University of Salamanca. He serves as the Principal Investigator of the EMC3 research group. His main research interests include Climate Change Education and ultrafast laser interactions.

Chapter 7
Exploring the Link Between Sustainability Success and Student Knowledge and Attitudes: TIMSS 2023 Perspectives

Svatava Janoušková, Paulina Koršňáková, Libor Klement, Michaela Janoušková, Michal Blaško, and Pavel Zikl

7.1 Introduction

As stated in a vast body of literature, our planet faces significant challenges, including climate change, habitat and ecosystem destruction, and pollution of air, water, and soil caused by human activities (see, for example, Richardson et al., 2023; Rockström et al., 2009). Long-term, unprecedented environmental

S. Janoušková (✉)
Faculty of Science, Charles University, Prague, Czech Republic
e-mail: svatava.janouskova@natur.cuni.cz

P. Koršňáková
International Association for Evaluation of Educational Achievements (IEA), Amsterdam, The Netherlands
e-mail: p.korsnakova@iea.nl

L. Klement
Faculty of Education, University of Hradec Králové, Hradec Králové, Czech Republic

Czech School Inspectorate, Prague, Czech Republic
e-mail: libor.klement@csicr.cz

M. Janoušková
Faculty of Education, University of South Bohemia in České Budějovice, České Budějovice, Czech Republic
e-mail: Janouml8@jcu.cz

M. Blaško
Faculty of Science, Charles University, Prague, Czech Republic

Czech School Inspectorate, Prague, Czech Republic
e-mail: blaskomi@natur.cuni.cz

P. Zikl
Faculty of Education, University of Hradec Králové, Hradec Králové, Czech Republic
e-mail: Pavel.zikl@uhk.cz

D. Hastedt et al. (eds.), *Educating Pupils for Environmental Sustainability in Europe*, IEA Research for Education 18,
https://doi.org/10.1007/978-3-032-14725-7_7

degradation has led to the development of the concept of *sustainable development*, aimed at ensuring the protection of Earth's resources and ecosystems for the next generations.

In 2015, the United Nations established the Sustainable Development Goals (SDGs) to create a universal framework that addresses the world's most pressing challenges (United Nations, 2015). The SDGs serve as a blueprint for governments, businesses, and civil society to align and monitor their policies and actions toward common goals. This alignment helps maximize the impact of efforts and resources, fostering collaboration and innovation across sectors.

Education for sustainable development is embedded into SDG 4 to foster a more aware and capable generation that can drive innovation, promote responsible citizenship, and build a sustainable and resilient future for all. Considering this, it is unsurprising that efforts are being made at local, national, and international levels to identify the knowledge, skills, values, and attitudes that could enable students to contribute to achieving the SDGs.

The Trends in International Mathematics and Science Study (TIMSS) 2023, in addition to testing students' knowledge in key cognitive domains, also introduced questionnaires that focus on students' environmental attitudes and behaviors. The inclusion of these affective and behavioral questions provides a unique opportunity to study students' environmental knowledge, values, and attitudes together with other information collected about participating countries[1] and their progress toward sustainability.

In this chapter, the complex picture of seven culturally different European countries is presented: Albania, Austria, Bosnia and Herzegovina, the Czech Republic, Hungary, the Slovak Republic, and Sweden (Fig. 7.1). To do this, information from TIMSS 2023 is reviewed alongside information obtained from the Sustainable Development Report 2024 (Sachs et al., 2024).[2] The Sustainable Development Report 2024 provides additional details on responsible production and consumption, waste and wastewater management, biodiversity protection, and other environmental sustainability issues in the communities of these seven countries, offering a more accurate picture of the context in which students live and the factors that may shape their worldviews (including their knowledge, values, and attitudes). Building on the assumption that students' worldviews are shaped by the wider community and the environment in which they live, it enables a more in-depth understanding of results related to the TIMSS 2023 Environmental Attitudes and Behaviors Framework.

[1] Although TIMSS, and this volume, refers to participants as *education systems*, in this chapter the terms *country/countries* are used because much of the chapter deals with national-level social, economic, and environmental indicators.

[2] Since 2016, the global Sustainable Development Report has offered the latest data to monitor and compare the progress of all United Nations Member States toward achieving the SDGs. The report is based on data collected by leading international organizations and is prepared by a group of experts in the field of sustainable development. This report serves as a basis for discussions on sustainable development at both political and scientific levels.

Fig. 7.1 Countries included in the present study

The following research questions guide the analysis in this chapter:

1. What patterns of knowledge and attitudes can be identified among students in the seven analyzed countries, based on the TIMSS 2023 data?
2. What patterns in community environmental sustainability achievements (for example, related to waste and wastewater management, climate action, biodiversity protection) can be identified across the seven analyzed countries that may affect students' knowledge and attitudes?

7.2 Study Background

7.2.1 Sustainability Worldview

Sustainable development is a notion widely accepted as necessary for the development of society. It can be seen as a shared *worldview* construct that embraces a diversity of views in collaborative research and the co-production of knowledge related to sustainable development (de Vries, 2019; Van Opstal & Hugé, 2013). De Vries describes the need to balance the individual and society and to balance the material with the immaterial to achieve such a shared worldview (de Vries, 2019). It requires finding a balance between meeting the needs of one's personal life within a family or community and the responsibility to share in the broader development of a whole society. This also involves balancing material needs—as sources of comfort and pleasure—with spiritual values, which can offer meaning that excessive material wealth may fail to provide. Additionally, Van Opstal and Hugé (2013) note that people need cognitive, perceptual, and affective maps to orient themselves toward sustainable development.

Hedlund-de Witt (2012) describes worldview as a system of five overlapping dimensions: ontological (basic presumptions on the nature of reality), epistemological (modes of knowledge), axiological (particular values), anthropological (the nature of human beings and their position in the universe), and societal (how society is organized). This definition aligns with other conceptualizations of worldview found in the literature (for example, Koltko-Rivera, 2004; Rigolot, 2018).

Given that worldview is a multidimensional concept, answering the research questions guiding this chapter requires an understanding of what knowledge is, what shapes human values and attitudes, and the influence of society. Therefore, how knowledge, values and attitudes, and *societal context*—concepts which do not have universally accepted definitions—are understood in the present chapter is addressed. From a methodological perspective, this clarification is also essential, as it justifies the selection of the sources used for the descriptive analysis.

Just as the definition of worldview is complicated, so is the definition of knowledge. Some authors see knowledge as justified true belief (Janoušková et al., 2020), while others see it as an idea believed or entertained as plausible by a person regardless of its objective correctness, or a multi-polar paradigm composed of rational, emotional, and spiritual dimensions of knowledge (Bolisani & Bratianu, 2018; Taber, 2019). In the present chapter, knowledge is understood as well-understood (scientific) concepts that correspond with the current understanding of existing facts (objects, phenomena, and processes) (Janoušková et al., 2020). It is recognized that knowledge emerges through an ontological process, in which individuals construct understanding by evaluating both their own presumptions and the perspectives of others (see, for example, Rosida et al., 2023). However, to achieve sustainability goals, people must make informed and responsible decisions. Therefore, knowledge must be based on a proven theory, principle, or law. Incorrect knowledge misleads the public and policymakers, causing poor decisions that undermine genuine sustainability efforts. This diverts resources from effective solutions and undermines public trust, ultimately slowing progress towards sustainability goals.

Another important worldview dimension is the axiological dimension, which represents particular values. According to Rokeach (1968, 1973) and Schwartz and Bilsky (1987), values guide, motivate, and influence attitudes and behaviors. Values determine what people attend to, how they evaluate various aspects of the situation, and what alternatives they consider, affecting their actual behavior.

According to psychologists, there are several types of values (De Groot & Steg, 2008; Schultz, 2002; Steg & De Groot, 2012). In this study, environmental values are the central focus. Values related to pro-environmental beliefs, norms, intentions, and actions are divided into three categories: *egoistic values*, leading persons to act in their interest; *altruistic values*, reflecting a concern with the welfare of others;

and *biospheric values*, reflecting a key problem with the quality of nature and the environment (De Groot & Steg, 2008; Schultz, 2002; Steg & De Groot, 2012). Since environmental values reflect a form of self–nature connection—referring to a deep sense of connection to the natural environment and the belief that nature is a fundamental part of individual identity (see, for example, Martin & Czellar, 2017)—they encompass not only the axiological but also the anthropological dimensions of worldview.

Several studies have explored the relationships between these values and pro-environmental behavior. In literature, altruistic and biospheric values give rise to pro-environmental behavior, while egoistic values are often suggested to result in resistance or active opposition to environmental protection (Snelgar, 2006). Other researchers report that the psychological distance of environmental problems (such as climate change and local pollution) moderates the relationship between egoistic values and pro-environmental attitudes and behavior (Lou et al., 2024).

Although some authors state that values influence behavior indirectly via behavior-specific beliefs, preferences, attitudes, and norms (Steg & De Groot, 2012), it makes sense that there are attempts to measure them. By understanding values, it can be better predicted how individuals and groups might act in different situations, especially around issues like sustainability. This insight helps design interventions and policies that align with people's core motivations, making positive behavior change more likely.

Additionally, societal context (including, for example, family, community, and social media) plays a crucial role in worldview development. It affects understandings of the dynamic world and moral standards (Ajzen, 1991; Chwialkowska et al., 2020; Koltko-Rivera, 2004; Quimby & Angelique, 2011; Van Opstal & Hugé, 2013). Families are primary agents of socialization, instilling knowledge and values related to environmental care and sustainability in children and adolescents (Palacios Madero et al., 2023). Later, peers and other community members provide a sense of belonging and help shape social identity. This interaction is neither fully private nor public (Gilchrist, 2013). Social media has emerged in recent years as a new, significant aspect of societal context with social media platforms creating global virtual communities (see, for example, Swigger, 2013).

According to van Egmond and de Vries (2011), value orientation in society can be divided into two dimensions, contrasting between *individual* versus *collective* and *im-materialistic* (religious) versus *materialistic* (worldly). In their view, "Development is sustainable if some balance between material and immaterial and between individualist and collective values can be maintained. In this definition, sustainable development is identical to maintaining 'human dignity,' where human dignity coincides with the adjusted definition of the integral worldview" (van Egmond & de Vries, 2011, p. 866). However, finding such a value balance in democratic societal systems is as tricky as defining the exact content of the concept of sustainable development.

TIMSS presents a unique opportunity to investigate students' knowledge, values, and attitudes toward the natural environment. It can explain how family, school, community, educational curricula, and societal attitudes and actions toward environmental issues influence students. Although several studies have shown that there is no straightforward, linear relationship between environmental knowledge, values, attitudes, and actual pro-environmental behavior, this does not diminish the importance of examining these factors (see, for example, Pe'er et al., 2007).

7.2.1.1 Environmental Knowledge Tested in TIMSS 2023

Since environmental topics have been included in most countries' curricula, these topics have been a part of the TIMSS survey since 1995 and are outlined in the TIMSS science framework. The TIMSS 2023 science framework, analyzed in this study, is structured around content domains. For the fourth grade, the TIMSS 2023 science framework establishes three content domains: life science, physical science, and Earth science. For the eighth grade, four content domains are determined: biology, chemistry, physics, and Earth science (Centurino & Kelly, 2021).

Each of these content domains includes several major topic areas, and each topic area in turn contains one or more topics. Each topic is further described by specific objectives that represent the students' expected knowledge, abilities, and skills assessed within each topic (Centurino & Kelly, 2021).

Environmental topics are covered in the topic area of ecosystems. In this topic area, fourth-grade students must demonstrate basic knowledge of communities, food relationships (for example, distinguish between predator and prey), and competition. Environmental topics are also represented in the topic area of organisms, environment, and their interaction. In this topic area, students should show their knowledge of the positive and negative impacts of human activity on the essential components of the environment (water, soil, air). Students should also be able to provide general descriptions and examples of the effects of pollution on humans, plants, and animals. Environmental topics are also covered in the topic areas of forms of energy and energy transfer, and Earth's physical characteristics, resources and history. Here, students should be able to identify energy sources, including renewable sources, and their responsible use in everyday life. Environmental topics are also included in the topic area of Earth's weather and climates. The content of this area specifically focuses on students' ability to explain how changes in the Earth's average temperature in recent periods have impacted its physical properties. For example, ocean levels have risen, glaciers have melted, rivers have dried up, and deserts have expanded (see Centurino & Kelly, 2021; von Davier, Kennedy, et al., 2024).

Environmental topics are also a part of eighth-grade curricula. It mirrors the TIMSS fourth-grade students' survey. Here, students must demonstrate their understanding of the concept of ecosystems (such as, the flow of energy in ecosystems, relationships among populations of organisms in an ecosystem, factors affecting population size in an ecosystem, and human impact on the environment). Environmental topics are also covered in the topic area of Earth's processes, cycles, and history, where students should be able to demonstrate knowledge, for example, of how the environment has changed over long periods, give evidence for climate changes, or demonstrate the ability to interpret data or maps of weather patterns. Environmental topics are also included in the topic area of Earth's resources, their use and conservation. This area includes topics such as natural resource management, focusing on water resources and land use. It emphasizes students' understanding of renewable and non-renewable energy sources, including their advantages and disadvantages, their knowledge of the waste management hierarchy, and their awareness of the sustainable use of water resources and landscapes. Environmental topics are also included in the topic area of Earth's structure and physical features, with a focus on the importance and availability of freshwater resources, as well as the carbon cycle (see Centurino & Kelly, 2021; von Davier, Kennedy, et al., 2024).

7.2.1.2 Environmental Attitudes in TIMSS 2023

The above-described knowledge and abilities cover what needs to be known and understood to support the achievement of the SDGs in an informed way. Due to the increasing urgency to achieve the SDGs, TIMSS 2023 used student questionnaires that measure fourth- and eighth-grade students' attitudes toward the natural environment and their enactment of environmentally responsible behaviors. TIMSS 2023 draws from the two-dimensional Model of Ecological Values to measure students' environmental attitudes. This model is based on the Theory of Ecological Attitude (Reynolds & Komakhidze, 2022). This theory distinguishes two types of attitudes: preservation and utilization. Preservation reflects an individual's endorsement of conservation and protection of nature. Utilization demonstrates an individual's endorsement of human use of natural resources and the environment (Reynolds & Komakhidze, 2022). Reynolds and Komakhidze (2022) build on the theory of Wiseman and Bogner (2003), which sees individuals with *utilization attitudes* as those who see nature as serving human purposes and place trust in human innovation and progress to address environmental challenges. In contrast, individuals aligned with *preservation attitudes* will likely value spending time in nature and prioritize protecting natural areas.

The TIMSS 2023 survey of students' pro-environmental values was conducted using context questionnaires.[3] Students were scored based on how much they agreed with six statements about environmental preservation, using the Students Value Environmental Preservation scale (see Fig. 7.2). The scale is divided into three categories. Students who *very strongly value* environmental preservation scored at or above the cutoff, meaning they mostly agreed with three statements and slightly agreed with the other three. Students who *somewhat value* environmental preservation scored at or below the cutoff, meaning they slightly agreed with three

[3] The TIMSS 2023 context questionnaire scales are created using the Rasch partial credit model (PCM). Scores calculated using the PCM calibration are converted to a scale with an average of 10 and a standard deviation of two. This scale is divided into high, medium, and low regions using two cutoff points. Each region is linked to specific content interpretations based on the most likely response categories associated with those cutoff points (von Davier, Fishbein, et al., 2024).

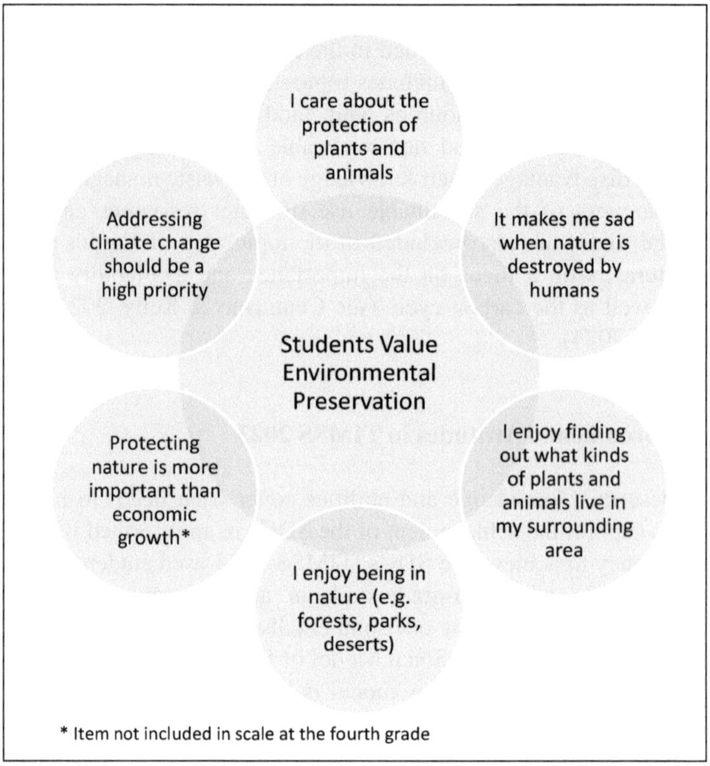

Fig. 7.2 Students value environmental preservation scale

statements and slightly disagreed with the other three. The remaining students *strongly value* environmental preservation (von Davier, Kennedy, et al., 2024).

7.2.2 Information Provided by SDGs Indicators as Context for Worldview

As stated, worldview combines knowledge, values, and societal context. TIMSS 2023 explores students' understanding of environmental concepts and identifies their pro-environmental attitudes. The focus is therefore on the environmental pillar—that is, environmental sustainability—within the broader concept of sustainable development. The extent to which environment (the place people live) influences attitudes toward that environment must also be explored. Many academic publications emphasize that both the social and natural environment shape knowledge, attitudes, and values, and thus, worldview.

One crucial aspect shaping people's worldview is direct experience with nature. People living in rural or pristine natural environments may develop a deeper connection to nature through direct interaction. Berenguer et al. (2005) showed that urban residents adopt more values of environmental responsibility but exhibit fewer pro-environmental attitudes and behavioral intentions. In contrast, rural residents demonstrate stronger attitudes toward environmental responsibility and greater consistency in their intentions to protect the environment. More recent studies (for example, Garner, 2017; Goula et al., 2015) have not found such clear differences. However, differences exist; for example, the attractiveness of certain environmental elements varies between urban- and rural-originated students, depending on feelings of familiarity or the need to escape from anxiety or monotony (Goula et al., 2015). In many countries, the distinction between rural and urban areas is blurred, partly because people commute to cities for work and, conversely, people from cities have relatively easy access to nature outside the city (Garner, 2017; Goula et al., 2015).

There are countries, however, where the differences in pro-environmental attitudes between people in rural and urban areas are pronounced, partly due to the remoteness of many rural areas. In the People's Republic of China, rural residents are less concerned about the environment than their urban counterparts, especially with respect to pollution, nature conservation, and global environmental degradation. At the same time, rural residents tend to focus more on problems related to agricultural production (for example, land erosion or desertification) (Yu, 2014).

Studies generally show that attitudes are shaped by the problems of the surrounding area, such as air quality or noise in cities, land degradation in rural areas, and overall access to information. Cultural, spiritual, and socioeconomic status factors play a significant role (Franzen & Vogl, 2013; Ignatow, 2006). The reasons for certain attitudes towards environmental sustainability may be egoistic, altruistic, or biospheric. However, they usually reflect the actual or foreseeable environmental problems individuals experience or fear (Schultz, 2002).

Assuming that the state of the environment influences, to some extent, attitudes towards it, then the state of the environment in each country provides important context for understanding how students' values and attitudes are formed. International reports that review progress toward environmental sustainability using carefully selected environmental indicators allow for comparable information across countries. Since all of the countries analyzed in this chapter are part of SDG progress assessment, the 2024 Sustainable Development Report (Sachs et al., 2024) is particularly useful, as it includes such indicators and provides information on trends in achieving these goals for the analyzed countries.

7.3 Methodology

To identify patterns of knowledge and attitudes based on TIMSS 2023 data (addressing the first research question), and patterns linked to community environmental sustainability achievements (that may affect students' knowledge and attitudes) (addressing the second research question) seven countries are examined in this analysis: Albania, Austria, Bosnia and Herzegovina, the Czech Republic, Hungary, the Slovak Republic, and Sweden.

7.3.1 Country Selection

The fundamental differences between these countries allow observation of whether variables such as students' knowledge, values and attitudes differ according to socioeconomic levels or environmental sustainability achievements.

Austria and Sweden, with strong pro-environmental policies and high rankings in the Sustainable Development Report 2024 (Sachs et al., 2024), serve as benchmarks for effective sustainability efforts. Despite following European Union policies, the Czech Republic, Hungary, and the Slovak Republic face some public skepticism about climate change, with ongoing debate about prioritizing economic benefits over environmental protection. Albania and Bosnia and Herzegovina, with weaker policies and lower rankings in the Sustainable Development Report 2024, reflect challenges tied to socioeconomic constraints and governance.

Including TIMSS 2023 rankings adds another layer to the analysis, as these rankings reflect differences in education systems and student achievement in science, which is closely linked to environmental knowledge.

The selected countries, with differing results in TIMSS, offer valuable insights into the relationship between knowledge, pro-environmental values, and the unique national contexts that approach environmental issues in varied ways. These differences help identify patterns in how education quality and achievement influence students' environmental awareness and attitudes. Additionally, the analysis reveals how national perceptions of the urgency to address environmental challenges shape pro-environmental values. By comparing countries with contrasting approaches to sustainability, the study uncovers the interplay between education, values, and societal priorities in addressing environmental problems.

7.3.2 Method and Variable Selection

Descriptive analysis is used to answer the research questions. Patterns in worldviews or environmental perceptions are context-dependent, as outlined in the theoretical part of this study. The descriptive method enables identification of initial trends or relationships in knowledge, values, and sustainability achievements which can later be tested with inferential statistics.

Emphasis is placed on selecting appropriate variables. TIMSS provides robust data on students' environmental knowledge and attitudes. As such, data on students' knowledge were taken from TIMSS 2023 environmental items and data on students' attitudes and behavior from the TIMSS 2023 context questionnaire.

Several data sources contribute to SDG indicators and can be used to describe environmental sustainability. These include the Global Indicator Framework for the Sustainable Development Goals (United Nations, 2017), indicators provided by the SDG Tracker from Our World in Data (2024), the Sustainable Development Report (Sachs et al., 2024), and Eurostat's report on the European Union's progress toward the SDGs (European Union, 2023).

For this study, a group of experts in environmental sustainability indicators were asked to analyze the available indicator sets. The goal of this analysis was to select an appropriate indicator set that provides comprehensive information on the environmental sustainability of the countries analyzed. Experts selected the Sustainable Development Report (Sachs et al., 2024) because it transforms indicators into values that enable straightforward and meaningful comparisons across the evaluated countries. The report also demonstrates more clearly the trends in achieving targets within countries. At the same time, the environmental indicators provide a good description of any environmental problems identified in the countries that could shape environmental attitudes (see Appendix 7.1).

7.3.3 Additional Information for Analysis

As outlined in the study background, worldview related to sustainability is a multidimensional concept influenced by a range of factors. In addition to insights into students' knowledge, values, and attitudes, and the environmental context of each country, data on economic and social factors are considered equally important (these are presented in the country profiles in Sect. 4). This perspective aligns with recommendations from an expert group to include social or economic context in cases where social and economic sustainability may conflict with environmental sustainability.

7.4 Country Profiles: Overall Information and SDGs Achievement

Country profiles offer a vital context for interpreting survey results by connecting students' environmental knowledge and attitudes to their countries' broader socioeconomic and environmental conditions. By highlighting each country's unique economic and ecological realities, these profiles allow for a more comprehensive understanding of how national contexts shape students' perspectives and awareness of environmental issues.

Additionally, the profiles reveal disparities between countries, shedding light on why students from different regions may display varying levels of environmental awareness and varying degrees of pro-environmental values. Including country-specific data ensures a nuanced analysis that avoids overgeneralization, respects differences, and accounts for the diverse challenges and priorities that influence global environmental knowledge and attitudes.

The following section presents the profiles of the individual countries included in the analysis, summarizing key economic, social, and environmental indicators. Supplementary data on environmental quality in each country is provided in Appendix 7.1.

As noted, the analysis compares countries with contrasting approaches to sustainability. The order of country profiles therefore follows the groupings outlined in Sect. 3.1 on the selection of the countries.

7.4.1 Albania

Albania is located in Southeastern Europe and has a population of approximately 2.79 million. As of 2023, the services sector constitutes the largest portion of Albania's gross domestic product (GDP), accounting for approximately 48 percent of the total. This sector encompasses various industries, including tourism, banking, telecommunications, and retail. Tourism, in particular, has been a significant contributor, with over 6.4 million tourists visiting Albania in 2019. The industrial sector represents about 21 percent of GDP, while agriculture accounts for approximately 18 percent. Albania is rich in natural resources (chromium, copper, nickel) and has oil and gas (Patos-Marinza oil field is one of Europe's largest onshore oil fields). Albania's primary energy source is hydropower (Central Intelligence Agency [CIA], 2024a).

In the overall ranking of 167 countries included in the Sustainable Development Report 2024 (Sachs et al., 2024), **Albania ranks 42nd**. According to the report, major challenges remain in wastewater treatment (water pollution from industrial and domestic effluents) and ocean protection. Although the fisheries sector does not contribute significantly to GDP (approximately 0.3%), the long coastline along the Adriatic and Ionian Seas provides excellent conditions for marine fishing. While the fisheries sector is economically relatively small, it is important from a sociopolitical point of view because it is often located in areas of high unemployment and socio-economic needs (Eurofish, 2023). Fishing is a sensitive issue linking environmental and social sustainability.

Terrestrial and coastal ecosystem biodiversity is at risk. Given that Albania is a tourist destination for both summer and winter recreation, tourism must be developed following the principles of environmental sustainability to avoid disturbing biodiversity (Vejsiu & Stafasani, 2017). The mining industry can also have adverse effects on biodiversity. Ensuring the protection of terrestrial ecosystems remains one of the country's significant environmental challenges (CIA, 2024a).

Air pollution and access to improved water sources are also significant challenges. All of these major challenges can affect people's attitudes towards the environment (CIA, 2024a; Sachs et al., 2024).

7.4.2 *Bosnia and Herzegovina*

Bosnia and Herzegovina, situated in Southeastern Europe, has a population of 3.12 million people. Estimates from 2023 show that the services sector (retail trade and restaurants) constitutes the largest portion of Bosnia and Herzegovina's GDP, accounting for approximately 56 percent of the total (CIA, 2024b). The economy also relies on industry—mining and the export of metals (steel, coal, iron ore, lead, zinc, manganese, bauxite), textiles, tobacco products, and wooden furniture (23% of the total GDP). Agriculture accounts for approximately 18 percent of the GDP, but one third of its land is cultivated or in pasture (CIA, 2024b; International Trade Administration, 2024).

Bosnia and Herzegovina ranks 50th among the 167 countries in the Sustainable Development Report 2024 (Sachs et al., 2024). Among the major challenges is inadequate anthropogenic wastewater treatment and limited sites for disposing of urban waste. On the other hand, the production of municipal solid waste (kg/capita/day) meets SDGs (CIA, 2024b).

Bosnia and Herzegovina also faces major challenges in decreasing CO_2 emissions from fossil fuel combustion. Most CO_2 emissions are in the energy sector, and coal is the predominant fossil fuel burned (International Trade Administration, 2024). Burning fossil fuels and other industrial activities are associated with poor air quality, reflected in elevated concentrations of $PM_{2.5}$. Air quality remains one of the major challenges for the country (CIA, 2024b).

Bosnia and Herzegovina, like Albania, faces a significant challenge in protecting terrestrial areas that are important for biodiversity. Bosnia and Herzegovina is part of the Dinaric Alps, one of Europe's biodiversity hotspots. The country harbors numerous endemic species and diverse ecosystems. For now, the government's focus is on improving people's quality of life. Industrial development, resource extraction, intensive agriculture, and urbanization are seen as one way to achieve this. This creates conflicts between economic needs and biodiversity protection (CIA, 2024b).

7.4.3 *Czech Republic*

The Czech Republic is situated in Central Europe and has a population of 10.89 million. As of 2023, the services sector constitutes the largest portion of the Czech Republic's GDP, accounting for approximately 59.8 percent of the economy. These services reflect the Czech Republic's position as a modern, developed economy with a strong emphasis on trade, tourism, and innovation; an even larger share of the service sector in GDP is projected in the coming years. The industrial sector follows in importance, accounting for about 30.2 percent of GDP. Key industries include automotive manufacturing, engineering, and electronics. The agriculture sector represents a smaller share, around 1.6 percent of GDP, and the agricultural land area continues to decline (CIA, 2024c).

The **Czech Republic ranks 12th** among the 167 countries in the Sustainable Development Report 2024 (Sachs et al., 2024). One of the major challenges of the Czech Republic is its high carbon footprint, which is caused by its heavy reliance on coal and natural gas for energy and heating, as well as energy-intensive industries. Transportation emissions and a slower transition to renewable energy also play significant roles. The quality of air and waste production remains a challenge. However, in the last 35 years, the environment has significantly improved, particularly in air and water quality (Cenia, 2008; Cenia, 2022); some of these environmental improvements are shown in Appendix 7.1.

7.4.4 Hungary

Hungary is situated in Central Europe and has a population of 9.59 million. The services sector plays a significant role in Hungary's economy, followed by industry, with agriculture contributing a smaller portion. Agriculture accounts for approximately 4.7 percent of GDP, industry for 24.3 percent, and the services sector makes up the remainder. The automotive sector is particularly prominent in the industry. Other important industries include electronics, pharmaceuticals, and food processing. Although the share of agriculture is very low, land used for farming covers 53 percent of the country's territory (CIA, 2024d).

Hungary ranks 20th among the 167 countries in the Sustainable Development Report 2024 (Sachs et al., 2024). Its relatively high carbon footprint is primarily due to its reliance on fossil fuels for energy production. Hungary's strong industrial sector, particularly energy-intensive industries like automotive manufacturing, contributes significantly to greenhouse gas emissions. CO_2 emissions remain a major challenge. Like the Czech Republic and the Slovak Republic, the air quality in Hungary does not meet SDG standards, as indicated by the annual mean of $PM_{2.5}$ concentration. While Hungary has a network of national parks and protected areas, managing terrestrial sites that are important to biodiversity remains a challenge (Csákvári et al., 2021; Government of Hungary, 2023).

7.4.5 Slovak Republic

The Slovak Republic is a country in Central Europe with 5.42 million inhabitants. The service sector drives GDP, accounting for 56.5 percent of the economy. The industrial sector focuses on automotive manufacturing. The country is notable for being the world's largest per capita car producer, with significant automobile assembly plants. The industrial sector accounts for 32.7 percent of GDP, bolstered by foreign IT, telecommunications, and outsourcing investment. Agriculture makes up a smaller share, at two percent (CIA, 2024e).

The Slovak Republic ranks 21st among the 167 countries in the Sustainable Development Report 2024 (Sachs et al., 2024). Similar to the Czech Republic, the Slovak Republic's major challenge is its high carbon footprint. The country relies significantly on fossil fuels, particularly coal and natural gas, for electricity and heat generation. Hard energy-intensive industries, particularly steel production and automotive manufacturing, contribute significantly to the country's carbon emissions (CIA, 2024e). The quality of air and access to improved water sources remain challenges. In the Slovak Republic, access to piped water is particularly difficult in rural areas, due to historical underinvestment, geographical obstacles, and high costs of extending infrastructure. Air quality is a problem due to emissions from industrial activities, domestic heating with coal and wood, and vehicle pollution, particularly in urban areas (Organisation for Economic Co-operation and Development [OECD], 2025).

7.4.6 Austria

Austria is a largely mountainous landlocked country in South-Central Europe with a population of 9.11 million. Austria's economy is primarily driven by the services sector, which accounts for over 70 percent of its GDP. Industry contributes around 28 percent, with strong sectors like machinery, vehicles, and chemicals. Agriculture is a minor industry, making up about one percent of the GDP. This economic structure reflects Austria's focus on advanced industries and a well-developed service economy (CIA, 2024f).

Austria ranks sixth among the 167 countries in the Sustainable Development Report 2024 (Sachs et al., 2024). Although Austria is renowned for its natural beauty and environmental awareness, like many developed nations it still faces its share of environmental challenges. As in other developed countries, CO_2 emissions remain a major challenge. CO_2 emissions are dominated by the burning of fossil fuels for energy production and industry. As in other developed countries, air quality remains a challenge; however, there is a positive trend toward better air quality. The recycling rate over the years 2015–2019 increased (European Environment Agency, 2022); however, the total amount of municipal waste also increased. Austria—along with Belgium, the Czech Republic, Denmark, Germany, Italy, Luxembourg, the Netherlands, and Slovenia—is one of the countries that is on track to meet 2025 targets in this area (European Environment Agency, 2022). Yet, like the Czech Republic, it is ranked in the Sustainable Development Report as a country with remaining challenges in the amount of non-recycled waste (Sachs et al., 2024).

Protected terrestrial areas in Austria are crucial for biodiversity but face challenges like habitat fragmentation, climate change, and conflicts with land use. Tourism and invasive species put additional pressure on ecosystems. The area protected in terrestrial and freshwater sites to maintain biodiversity remains a significant challenge. While the general public mostly enjoys the manifold benefits of conservation, such as the protection of landscapes and ecosystems and conservation of flagship species, and the contribution to the national natural and cultural heritage, the local population might stand disproportionately high costs in terms of restrictions to economic development (such as land use restrictions for commercial or residential purposes) (Withalm & Getzner, 2018).

7.4.7 Sweden

Sweden is a country located on the Scandinavian Peninsula in Northern Europe. Forests, lakes, and rivers dominate the landscape, but there is significant variation in landscape, temperature, and natural resources, which affects conditions for, for example, industry and agriculture. As of 2022, Sweden's GDP was comprised of approximately 1.5 percent from agriculture, 24 percent from industry, and 63 percent from services. This distribution reflects Sweden's advanced economy, emphasizing services and industrial production (CIA, 2024g).

Sweden ranks second among the 167 countries in the Sustainable Development Report 2024 (Sachs et al., 2024). Despite being one of the top ranked countries in the report, Sweden faces several environmental challenges. The major challenges relate to biodiversity, specifically protected areas in terrestrial, freshwater, and marine sites. The main reason is that traditional methods of farming and forestry, which once benefited many species, are now rarely used. Heavy nutrient loads and commercial fisheries adversely affect several marine and coastal environments (Naturvardsverket, 2024; Weitz et al., 2015). Sweden aims to integrate marine and water policies more effectively by balancing ecosystem protection with human needs, ensuring the preservation of biodiversity while maintaining a high quality of life for those relying on natural resources, such as fishers and coastal communities (Weitz et al., 2015). However, the process is relatively slow.

7.5 Country Profiles: TIMSS 2023 Environmental Awareness Results

In this section, TIMSS 2023 results on the environmental knowledge of grade four and grade eight students are presented.[4] It is important to note that each topic area includes a different number of items in the TIMSS test booklets, and the difficulty of the items also varies.[5]

[4] Not all countries participated in TIMSS 2023 with both grades. Albania, Bosnia and Herzegovina, and the Slovak Republic participated only for grade four. Austria participated only for grade eight. The Czech Republic, Hungary, and Sweden participated in both grade four and grade eight.

[5] Information about the difficulty of TIMSS 2023 environmental items is provided in Appendix 7.2 and 7.3.

Despite these differences, reporting the percentage of students' success in each topic area within a country is still meaningful. It provides a basic overview of how students performed across the various components of the TIMSS test. These percentages can serve as performance benchmarks within a class, region or country and help identify students' strengths and weaknesses in relation to the curriculum. As such, they offer valuable input for further analysis and for considering potential instructional interventions. However, the results must be interpreted with regard to the structure and difficulty of the different test sections.

Where appropriate, students' scores on specific items are compared with those from other countries. Such comparisons are valid because the same items were used across all countries.

7.5.1 Albania

The average score for fourth-grade students in Albania in overall science was 491 points, while their average score in environmental knowledge was 483 points (Table 7.1). In both overall science and environmental knowledge, students' average scores reached the Intermediate International Benchmarks in TIMSS 2023. However, the average score in environmental knowledge is statistically significantly lower than their overall science score.

Students answered correctly on 73 percent of items in the topic area of forms of energy and energy transfer. This topic area is represented by just two items dealing with energy transfer in turbines. In contrast, the share of students with the correct responses was lower in items dealing with organisms, environment, and their interactions (44%) and Earth's physical characteristics, resources, and history (46%). The most difficult item for students was identifying the benefit of using sunlight or wind to produce electricity (Item ID SE71213): only 8.4 percent of students answered correctly. Although this item was so difficult that it was classified as too difficult to anchor, students from two other countries

Table 7.1 Science and environmental knowledge achievement (grade four)

	Average scale score
Average science achievement	491 (4.5)
Average environmental knowledge	483 (5.4)
Topic area	*Average percent correct across items*
Earth's physical characteristics, resources, and history	46
Earth's weather and climates	56
Ecosystems	49
Forms of energy and energy transfer	73
Organisms, environment, and their interactions	44

Note: Standard errors appear in parentheses

analyzed—Hungary and Sweden—achieved better results (41% and 24.6%, respectively). Another difficult item required identifying the ecosystem where living things are found (Item ID SE71080). Only 10.2 percent of students in Albania answered correctly, compared to higher shares in all other countries included in this analysis. However, it is important to note that this is a difficult item, classified at the advanced anchor level.

As in most other countries assessed in this study, the most substantial stated pro-environmental values in Albania are associated with high levels of environmental knowledge (Table 7.2). However, it is interesting to note that Albania's fourth graders have the highest proportion of students reporting to *very strongly value* attitudes toward environmental issues among all countries evaluated. In contrast, only about two percent of students report pro-environmental values and attitudes at a low level (*somewhat value*).

Most students in Albania are from lower or middle socioeconomic status groups (Table 7.3). Similar to other countries in this study, higher economic status corresponds with higher success rates in solving environmental items. However, students' pro-environmental attitudes as reflected in the Students Value Environmental Preservation scale are similar across all socioeconomic status groups. Likewise, the percentage of students who enjoy being in nature is similar across these groups. It is worth noting that, among the countries evaluated in this study, fourth graders in Albania have the highest proportion of students enjoying being in nature.

Table 7.2 Students valuing environmental preservation (grade four)

	Very strongly value	Strongly value	Somewhat value
Percent of students	77 (1.3)	21 (1.1)	2[a]
Average environmental knowledge *(average scale score)*	494 (5.5)	475 (8.5)	*393*

Notes: Standard errors appear in parentheses. [a] Although fewer than two percent of students in this region fall into this group, the average scale score is reported for information purposes

Table 7.3 Home socioeconomic status (grade four)

	Socioeconomic status		
	Higher	Middle	Lower
Percent of students	12 (1.7)	39 (2.9)	48 (3.4)
Average environmental knowledge *(average scale score)*	513 (12.8)	486 (8.6)	468 (7.8)
Students Value Environmental Preservation *(average scale score)*	11.2 (0.17)	11.2 (0.07)	10.8 (0.12)
I enjoy being in nature *(percent of students who agree a lot)*	84 (3.6)	86 (1.9)	82 (2.3)

Note: Standard errors appear in parentheses

Table 7.4 Science and environmental knowledge achievement (grade four)

	Average scale score
Average science achievement	448 (3.7)
Average environmental knowledge	446 (4.3)
Topic area	*Average percent correct across items*
Earth's physical characteristics, resources, and history	42
Earth's weather and climates	46
Ecosystems	41
Forms of energy and energy transfer	64
Organisms, environment, and their interactions	38

Note: Standard errors appear in parentheses

7.5.2 Bosnia and Herzegovina

Fourth graders in Bosnia and Herzegovina scored an average of 448 points in overall science, while their achievement in environmental knowledge was almost the same, with an average score of 446 points (Table 7.4). This corresponds to achievement at the Low International Benchmarks in TIMSS 2023 on both scales. In Bosnia and Herzegovina, 64 percent of students correctly answered items related to forms of energy and energy transfer. Challenging items were related to the topic areas of organisms, environment, and their interactions, and ecosystems. For example, a significant gap emerged on an item involving the relationship between two organisms in a food web, with only 7.1 percent answering correctly (Item ID SE81075). Among the countries analyzed, Sweden had the second-lowest score on this item (17.3%). However, this item is classified as advanced. Fourth graders from Bosnia and Herzegovina also scored relatively low on items related to human activities that help or harm the environment compared to the other countries analyzed. While more than 50 percent of students answered the item correctly across all assessed countries—and nearly 70 percent in Sweden—only 37 percent of students in Bosnia and Herzegovina succeeded (Item ID SE81055). However, this item was also anchored at a high level.

Bosnia and Herzegovina stands out among the countries examined in this study, as students' knowledge shows little variation regardless of whether they report to *very strongly value*, *strongly value*, or *somewhat value* environmental preservation (Table 7.5). However, 97 percent of fourth graders report *very strongly value* or *strongly value* on environmental preservation. This makes Bosnia and Herzegovina the second country, after Albania, with the lowest proportion of students reporting *somewhat value* on environmental preservation (3%).

The gap in environmental knowledge among students in Bosnia and Herzegovina varies considerably based on socioeconomic status (Table 7.6). There is a 96-point difference between students from the higher socioeconomic status group and those from the lower socioeconomic status group. Similar to Albania, there is no difference in the value placed on environmental preservation, as reflected in the Students

Table 7.5 Students valuing environmental preservation (grade four)

	Very strongly value	Strongly value	Somewhat value
Percent of students	61 (1.4)	36 (1.3)	3 (0.4)
Average environmental knowledge *(average scale score)*	452 (4.9)	451 (6.3)	454 (15.8)

Note: Standard errors appear in parentheses

Table 7.6 Home socioeconomic status (grade four)

	Socioeconomic status		
	Higher	Middle	Lower
Percent of students	20 (1.7)	48 (1.8)	32 (2.5)
Average environmental knowledge *(average scale score)*	504 (8.0)	463 (5.6)	408 (10.4)
Students Value Environmental Preservation *(average scale score)*	10.3 (0.13)	10.2 (0.07)	10.3 (0.08)
I enjoy being in nature *(percent of students who agree a lot)*	78 (1.7)	79 (1.8)	80 (3.5)

Note: Standard errors appear in parentheses

Table 7.7 Science and environmental knowledge achievement (grade four)

	Average scale score
Average science achievement	526 (2.3)
Average environmental knowledge	519 (3.1)
Topic area	*Average percent correct across items*
Earth's physical characteristics, resources, and history	48
Earth's weather and climates	62
Ecosystems	60
Forms of energy and energy transfer	82
Organisms, environment, and their interactions	49

Note: Standard errors appear in parentheses

Value Environmental Preservation scale, between these groups. Likewise, the preference for being in nature is the same across all socioeconomic groups.

7.5.3 *Czech Republic*

In both overall science and environmental knowledge, fourth-grade students' average scores meet the Intermediate International Benchmarks in TIMSS 2023. However, their achievement in environmental knowledge (average score 519) is statistically significantly lower than their overall science achievement (average score 526) (Table 7.7).

Fourth graders from the Czech Republic answered 82 percent of items correctly in the topic area of forms of energy and energy transfer. They outperformed students in all other countries analyzed in the topic area of ecosystems. They achieved 48 percent correct answers in the topic area of Earth's physical characteristics, resources, and history, and 49 percent in the topic area of organisms, environment, and their interactions.

Like their peers in Albania, Bosnia and Herzegovina, and the Slovak Republic, fourth graders in the Czech Republic also struggled with solving the item on the benefits of using sunlight or wind to produce electricity, with only 10 percent correct answers (Item ID SE71213). This outcome is surprising, especially compared to the significantly better achievement of students in Sweden and Hungary. Similar to students in other countries, they faced difficulties when solving the item about the relationship between damaged habitats and the populations of specific organisms (Item ID SE81054), with only 11.9% succeeding. However, it should be noted that this item was classified as too difficult to anchor. It is surprising that students from the Czech Republic performed relatively poorly in describing both the negative and positive impacts of human activities on the environment (Item ID SE81044); 75.4 percent of students succeeded, compared to 82.8 percent of students from Hungary and 90.6 percent of students from Sweden. This item was anchored at a low level.

Like fourth graders, eighth graders from the Czech Republic demonstrated statistically significantly weaker achievement in environmental knowledge than overall science achievement, with an average score of 517 points in environmental topics versus 527 points in overall science (Table 7.8). Eighth-grade students answered 60 percent of items correctly in the topic area of ecosystems. Eighth graders faced troubles when solving problems related to the topic area of Earth's processes, cycles, and history, performing significantly worse than students from Austria or Sweden in solving the items dealing with evidence for global climate change or the formation of natural resources. Eighth graders also struggled with solving problems in the topic area of Earth's resources, their use, and conservation, with only half of the students answering correctly. Compared to students in other countries analyzed in this study, students from the Czech Republic were notably less successful, particularly in solving the adverse effects of fertilizer (Item ID SE62235). However, this item is classified as advanced.

Table 7.8 Science and environmental knowledge achievement (grade eight)

	Average scale score
Average science achievement	527 (2.0)
Average environmental knowledge	517 (2.1)
Topic area	*Average percent correct across items*
Earth's processes, cycles, and history	45
Earth's resources, their use and conservation	50
Earth's structure and physical features	65
Ecosystems	60

Note: Standard errors appear in parentheses

Table 7.9 Students valuing environmental preservation (grades four and eight)*

	Very strongly value		Strongly value		Somewhat value	
	Grade 4	Grade 8	Grade 4	Grade 8	Grade 4	Grade 8
Percent of students	51 (1.0)	30 (0.7)	43 (0.9)	54 (0.7)	5 (0.3)	16 (0.5)
Average environmental knowledge *(average scale score)*	527 (3.5)	530 (2.7)	516 (3.2)	515 (2.3)	494 (6.2)	504 (3.7)

Notes: *A direct comparison between fourth and eighth graders is complicated, as the groupings are based on different cut-off points. Standard errors appear in parentheses

Table 7.10 Home socioeconomic status/home educational resources (grades four and eight)

	Home socioeconomic status/home educational resources					
	Higher/Many		Middle/Some		Lower/Few	
	Grade 4	Grade 8	Grade 4	Grade 8	Grade 4	Grade 8
Percent of students	47 (1.1)	32 (1.0)	48 (1.0)	53 (0.8)	5 (0.4)	15 (0.8)
Average environmental knowledge *(average scale score)*	553 (4.4)	558 (2.4)	509 (3.7)	510 (1.9)	462 (7.2)	456 (4.6)
Students Value Environmental Preservation *(average scale score)*	9.9 (0.05)	9.6 (0.04)	9.7 (0.05)	9.3 (0.03)	9.5 (0.12)	8.8 (0.05)
I enjoy being in nature *(percent of students who agree a lot)*	68 (0.9)	56 (1.0)	66 (1.2	46 (1.0)	62 (3.3)	39 (1.7)

Note: Standard errors appear in parentheses

When examining the pro-environmental values of fourth graders, approximately half of students report *very strongly value* (Table 7.9). Another 43 percent report *strongly value*, and only five percent exhibit low pro-environmental values (*somewhat value*). Among eighth graders, 30 percent of students report to *very strongly value* environmental preservation, while 54 percent report *strongly value*. In both fourth and eighth grades, students' average scores decline slightly, accompanied by a corresponding decrease in their concern for the environment.

The socioeconomic status of both fourth and eighth graders is linked to their level of environmental knowledge, with students from higher socioeconomic backgrounds demonstrating better environmental knowledge (Table 7.10). The Students Value Environmental Preservation scale scores, for students in both grades, follows a similar trend. However, in grade eight, there is a more pronounced difference between the students from higher and lower socioeconomic groups compared to grade four.

Grade-four students from the Czech Republic display similar pro-environmental values across all socioeconomic groups. The proportion of students who agree a lot that they enjoy spending time in nature is approximately the same across all three socioeconomic groups. The differences among socioeconomic groups are not statistically significant. Fourth graders enjoy spending time in nature significantly less than their peers in Albania and Bosnia and Herzegovina.

Eighth graders' preference for spending time in nature differs significantly from that of fourth graders. Eighth graders from the higher socioeconomic status group are less likely to spend time outdoors than fourth graders from the lower socioeconomic status group. The difference in eighth graders' preference for spending time in nature decreases as their socioeconomic status rises. The gap between groups is 10 and seven percentage points, respectively.

7.5.4 Hungary

Hungary stands out among the countries analyzed, as the achievement of fourth graders in environmental knowledge (average score 534) is statistically significantly higher than their science achievement (average score 524) (Table 7.11). In both overall science and environmental knowledge, students' average scores meet the Intermediate International Benchmarks in TIMSS 2023. Students achieved good results in the topic area of forms of energy and energy transfer, with 81 percent successfully solving the items. They also achieved good results in the topic area of Earth's weather and climates with 68 percent correct responses. The topic area of organisms, environment, and their interactions proved particularly challenging, with just 49 percent of students responding correctly. For example, within this topic area, students found it difficult to solve an item related to reducing air pollution in cities (Item ID SE81036), with only 26.8 percent achieving success. However, it should be emphasized again that this is an item considered too difficult to anchor.

Unlike fourth graders, eighth graders in Hungary have a slightly lower average score in environmental knowledge (518 points) compared to their overall science score (522 points) (Table 7.12). However, this difference is not statistically significant. In both overall science and environmental knowledge, students' average scores meet the Intermediate International Benchmarks in TIMSS 2023. Students faced

Table 7.11 Science and environmental knowledge achievement (grade four)

	Average scale score
Average science achievement	524 (3.2)
Average environmental knowledge	534 (3.8)
Topic area	*Average percent correct across items*
Earth's physical characteristics, resources, and history	55
Earth's weather and climates	68
Ecosystems	58
Forms of energy and energy transfer	81
Organisms, environment, and their interactions	49

Note: Standard errors appear in parentheses

Table 7.12 Science and environmental knowledge achievement (grade eight)

	Average scale score
Average science achievement	522 (3.3)
Average environmental knowledge	518 (3.7)
Topic area	*Average percent correct across items*
Earth's processes, cycles, and history	45
Earth's resources, their use and conservation	56
Earth's structure and physical features	53
Ecosystems	59

Note: Standard errors appear in parentheses

Table 7.13 Students valuing environmental preservation (grades four and eight)*

	Very strongly value		Strongly value		Somewhat value	
	Grade 4	Grade 8	Grade 4	Grade 8	Grade 4	Grade 8
Percent of students	59 (1.0)	43 (1.0)	36 (0.9)	46 (0.9)	5 (0.4)	11 (0.6)
Average environmental knowledge *(average scale score)*	549 (3.6)	534 (3.7)	522 (5.2)	515 (4.5)	483 (8.9)	483 (7.4)

Notes: * A direct comparison between fourth and eighth graders is complicated, as the groupings are based on different cut-off points. Standard errors appear in parentheses

significant difficulty with items in the topic area of Earth's processes, cycles, and history. As in other countries, the lowest success rate was observed in an item dealing with the impact of volcanic activity on the Earth's climate (Item ID SE82325). Eighth graders in Hungary had the lowest percentage of correct answers on evidence for global climate change (Item ID SE82327) among all the countries assessed, with only one third providing the correct answer. It should be mentioned again that the first item (Item ID SE82325) was classified as too difficult to anchor, and the second (Item ID SE82327) as a high-level item.

As in other countries analyzed, higher levels of knowledge are associated with stronger pro-environmental values in Hungary (Table 7.13). Among fourth graders, 59 percent report to *very strongly value* environmental preservation, a higher proportion than in the Czech Republic, the Slovak Republic, and Sweden. Among eighth graders, the proportion of students who reported to *very strongly value* environmental preservation decreases, while it increases for those who reported *strongly value* and *somewhat value*.

Students in both fourth and eighth grades from the higher socioeconomic status group demonstrate better environmental knowledge than those from the middle or lower socioeconomic status groups (Table 7.14). Among fourth graders, the gap between students from the higher and lower socioeconomic status groups is 135 points; a similar gap was found for eighth-grade students, at 137 points. Higher levels of knowledge are also connected to stronger pro-environmental values, as seen in the Students Value Environmental Preservation scale scores. As students

Table 7.14 Home socioeconomic status/home educational resources (grades four and eight)

	Home socioeconomic status/home educational resources					
	Higher/Many		Middle/Some		Lower/Few	
	Grade 4	Grade 8	Grade 4	Grade 8	Grade 4	Grade 8
Percent of students	47 (1.9)	42 (1.4)	40 (1.7)	41 (1.2)	13 (1.7)	17 (1.4)
Average environmental knowledge *(average scale score)*	584 (4.3)	565 (3.1)	536 (5.2)	508 (3.0)	449 (13.0)	428 (6.2)
Students Value Environmental Preservation *(average scale score)*	10.5 (0.05)	10.2 (0.05)	10.2 (0.07)	9.8 (0.04)	9.7 (0.19)	9.1 (0.08)
I enjoy being in nature *(percent of students who agree a lot)*	77 (1.1)	64 (1.3)	74 (1.8)	60 (1.0)	74 (2.9)	42 (2.4)

Note: Standard errors appear in parentheses

Table 7.15 Science and environmental knowledge achievement (grade four)

	Average scale score
Average science achievement	521 (3.3)
Average environmental knowledge	509 (4.4)
Topic area	*Average percent correct across items*
Earth's physical characteristics, resources, and history	45
Earth's weather and climates	60
Ecosystems	57
Forms of energy and energy transfer	79
Organisms, environment, and their interactions	46

Note: Standard errors appear in parentheses

reach grade eight, the preference for spending time in nature decreases. Eighth graders from higher socioeconomic backgrounds are likely to spend time outdoors, whereas those from lower socioeconomic backgrounds are less inclined to do so. The difference between these two groups is 22 percentage points.

7.5.5 *Slovak Republic*

Fourth graders in the Slovak Republic scored an average of 521 points in overall science, while their average score in environmental knowledge was 509 points (Table 7.15), meeting the Intermediate International Benchmarks in TIMSS 2023.

However, their achievement in environmental knowledge is statistically significantly lower than their overall science achievement. Fourth graders performed well in the topic area of forms of energy and energy transfer, achieving a success rate of 79 percent. Fourth graders found it most challenging to solve problems in the topic areas of Earth's physical characteristics, resources, and history, and organisms, environment, and their interactions. Fewer than a third of students could answer the question about how the farm impacts the fields in the near area, which was the second-lowest result among the countries assessed in this study (Item ID SE81076). Students also struggled to solve the question of reducing air pollution in a city (Item ID SE81036). Only 21.4 percent of students answered the item correctly, marking again the second-lowest result among the countries analyzed. However, it should be noted that the first item (Item ID SE81076) is classified as advanced, and the second (Item ID SE81036) as too difficult to anchor.

The pro-environmental values of fourth graders in the Slovak Republic closely resemble those of their peers in the Czech Republic. Students' environmental values vary, with 52 percent reporting *very strongly value*, 42 percent *strongly value*, and six percent *somewhat value* on environmental preservation (Table 7.16). As in the other countries, a connection is observed between the level of knowledge and the pro-environmental values declared. The higher the students' knowledge, the stronger their stated pro-environmental values.

The relationship between environmental knowledge and socioeconomic status among fourth graders in the Slovak Republic is consistent with findings from other countries. Students from the higher socioeconomic status group scored higher on item solving tasks (Table 7.17). It is worth noting that the knowledge gap between

Table 7.16 Students valuing environmental preservation (grade four)

	Very strongly value	Strongly value	Somewhat value
Percent of students	52 (1.3)	42 (1.1)	6 (0.5)
Average environmental knowledge *(average scale score)*	523 (4.9)	503 (5.5)	462 (12.5)

Note: Standard errors appear in parentheses

Table 7.17 Home socioeconomic status (grade four)

	Socioeconomic status		
	Higher	Middle	Lower
Percent of students	33 (1.3)	48 (1.2)	18 (1.2)
Average environmental knowledge *(average scale score)*	555 (4.0)	514 (4.3)	415 (11.2)
Students Value Environmental Preservation *(average scale score)*	9.9 (0.06)	9.9 (0.06)	9.2 (0.17)
I enjoy being in nature *(percent of students who agree a lot)*	70 (1.6)	77 (1.1)	69 (3.4)

Note: Standard errors appear in parentheses

students from the highest and lowest socioeconomic groups is 140 points in TIMSS 2023.

Students from high and middle socioeconomic groups express relatively high pro-environmental values (reported as *very strongly value*), as indicated by the Students Value Environmental Preservation scale. However, students from the lower socioeconomic status group report values of only 0.8 points lower, which is a positive finding. Interestingly, students from high and low socioeconomic groups enjoy spending time in nature equally. In contrast, students from the middle socioeconomic group tend to enjoy being in nature more than their peers from both high and low socioeconomic groups.

7.5.6 Austria

Only eighth-grade students participated in the TIMSS 2023 survey in Austria. However, this information is relevant to the study as Austria is one of the countries with a firm pro-environmental policy. Understanding the scientific and environmental knowledge and pro-environmental values that students in Austria possess is valuable.

The average science achievement of eighth-grade students was relatively higher than their average environmental knowledge (Table 7.18). In both overall science and environmental knowledge, students' average scores meet the Intermediate International Benchmarks in TIMSS 2023.

Students in Austria encountered certain difficulties when solving the item related to the effects of air pollution (Item ID SE82068B). The item was successfully solved by 26.8 percent of students. They also performed somewhat worse than other eighth-grade students included in this study on an item concerning the role of organisms in the energy pyramid (Item ID SQ82L01), with 54.4 percent correct answers. Both items fall under the topic area of ecosystems. On the other hand, students in Austria ranked among the best in the analyzed countries in more complex problems in this area requiring reasoning; for example, evaluating climate data or explaining high temperatures in city centers (Item ID SE72720). This item, categorized under

Table 7.18 Science and environmental knowledge achievement (grade eight)

	Average scale score
Average science achievement	512 (2.4)
Average environmental knowledge	501 (3.3)
Topic area	*Average percent correct across items*
Earth's processes, cycles, and history	48
Earth's resources, their use and conservation	50
Earth's structure and physical features	55
Ecosystems	53

Note: Standard errors appear in parentheses

Table 7.19 Students valuing environmental preservation (grade eight)

	Very strongly value	Strongly value	Somewhat value
Percent of students	33 (0.8)	49 (0.9)	18 (0.8)
Average environmental knowledge *(average scale score)*	526 (4.3)	502 (3.4)	464 (6.2)

Note: Standard errors appear in parentheses

Table 7.20 Home educational resources (grade eight)

	Home educational resources		
	Many	Some	Few
Percent of students	36 (1.1)	44 (0.9)	19 (1.0)
Average environmental knowledge *(average scale score)*	550 (3.0)	498 (3.3)	418 (6.8)
Students Value Environmental Preservation *(average scale score)*	9.8 (0.05)	9.3 (0.05)	8.9 (0.09)
I enjoy being in nature *(percent of students who agree a lot)*	51 (1.4)	45 (1.4)	36 (2.0)

Note: Standard errors appear in parentheses

the topic area of Earth's processes, cycles, and history, was classified as too difficult to anchor.

As in other countries, the level of science and environmental knowledge corresponds to the strength of pro-environmental values in Austria (Table 7.19). The expressed strength of pro-environmental values among eighth-grade students in Austria is comparable to that of their peers in the Czech Republic. One third of the students report to *very strongly value* environmental preservation, while the majority express *strongly value.* Nearly one fifth of the students report low pro-environmental values (*somewhat value*).

As with eighth graders in the Czech Republic and Hungary, socioeconomic status in Austria is reflected in home educational resources (Table 7.20). For students in Austria, higher socioeconomic status is associated with higher levels of environmental knowledge. Additionally, higher socioeconomic status increases the inclination to enjoy being in nature. Pro-environmental attitudes, as measured by the Students Value Environmental Preservation scale, decrease proportionally with socioeconomic status and environmental knowledge.

7.5.7 Sweden

In both overall science and environmental knowledge, fourth-grade students' average scores meet the Intermediate International Benchmarks in TIMSS 2023. However, their achievement in environmental knowledge (average score 526) is

Table 7.21 Science and environmental knowledge achievement (grade four)

	Average scale score
Average science achievement	533 (3.2)
Average environmental knowledge	526 (3.6)
Topic area	*Average percent correct across items*
Earth's physical characteristics, resources, and history	51
Earth's weather and climates	65
Ecosystems	57
Forms of energy and energy transfer	80
Organisms, environment, and their interactions	58

Note: Standard errors appear in parentheses

Table 7.22 Achievement and knowledge (grade eight)

	Average scale score
Average science achievement	521 (2.9)
Average environmental knowledge	534 (3.8)
Topic area	*Average percent correct across items*
Earth's processes, cycles, and history	53
Earth's resources, their use and conservation	53
Earth's structure and physical features	51
Ecosystems	64

Note: Standard errors appear in parentheses

statistically significantly lower than their overall science achievement (average score 533) (Table 7.21).

Similar to students in other countries included in this study, students in Sweden solved the items included in the topic area of forms of energy and energy transfer with high success. Students faced the most significant challenges in some items in the topic area of Earth's physical characteristics, resources, and history. Their achievement on an item about the disadvantages of farming near a river (Item ID SE81920) was the second lowest among the evaluated countries. However, this item was classified as advanced.

Unlike fourth graders, eighth graders have a significantly higher average score in environmental knowledge (534 points) compared to their overall science score (521 points) (Table 7.22). Eighth-grade students performed well on items in the topic area of ecosystems, achieving the best results among the countries analyzed in the study. Students in Sweden performed relatively well on challenging reasoning items, such as evaluating why climate data was incorrect (Item ID SE62211B). Despite only 11.4 percent answering correctly, they ranked among the most successful participants across the countries with eighth-grade data. This item was classified as too difficult to anchor.

Table 7.23 Students valuing environmental preservation (grades four and eight)*

	Very strongly value		Strongly value		Somewhat value	
	Grade 4	Grade 8	Grade 4	Grade 8	Grade 4	Grade 8
Percent of students	40 (1.0)	28 (1.1)	48 (0.9)	46 (0.9)	13 (0.6)	26 (1.0)
Average environmental knowledge *(average scale score)*	542 (4.7)	559 (4.7)	524 (4.6)	546 (4.7)	493 (6.3)	504 (5.1)

Notes: *A direct comparison between fourth and eighth graders is complicated, as the groupings are based on different cut-off points. Standard errors appear in parentheses

Table 7.24 Home socioeconomic status/home educational resources (grades four and eight)

	Home socioeconomic status/home educational resources					
	Higher/Many		Middle/Some		Lower/Few	
	Grade 4	Grade 8	Grade 4	Grade 8	Grade 4	Grade 8
Percent of students	55 (1.8)	42 (1.5)	39 (1.4)	43 (1.1)	6 (0.6)	15 (0.7)
Average environmental knowledge *(average scale score)*	565 (3.2)	592 (3.8)	502 (4.3)	511 (3.6)	449 (6.5)	449 (5.7)
Students Value Environmental Preservation *(average scale score)*	9.5 (0.06)	9.5 (0.06)	9.1 (0.05)	9.0 (0.05)	8.7 (0.17)	8.7 (0.10)
I enjoy being in nature *(percent of students who agree a lot)*	53 (1.4)	42 (1.4)	43 (1.7)	30 (1.3)	38 (3.4)	23 (1.9)

Note: Standard errors appear in parentheses

Students in Sweden demonstrate a clear link between environmental knowledge and pro-environmental values—the greater their knowledge, the stronger their values (Table 7.23). Eighth graders' pro-environmental values are similar to those of their peers in the other countries analyzed, while the proportion of fourth graders expressing to *very strongly value* environmental preservation is among the lowest.

In Sweden, students in both fourth and eighth grades from the higher socioeconomic status group demonstrate better environmental knowledge than those from the middle or lower socioeconomic status groups (Table 7.24). The gap between fourth graders from higher and lower socioeconomic groups is 116 points, while for eighth graders, it increases to 143 points—the largest difference observed among all surveyed countries. It is also worth noting that the proportion of children in the high and middle socioeconomic groups who like to spend time in nature is the lowest in all countries surveyed, among both grade four and eight students. Furthermore, their pro-environmental attitudes, as reflected in the Students Value Environmental Preservation scale, are somewhat lower than those in other countries. The scale values for eighth-grade students are approximately similar to those in the Czech Republic.

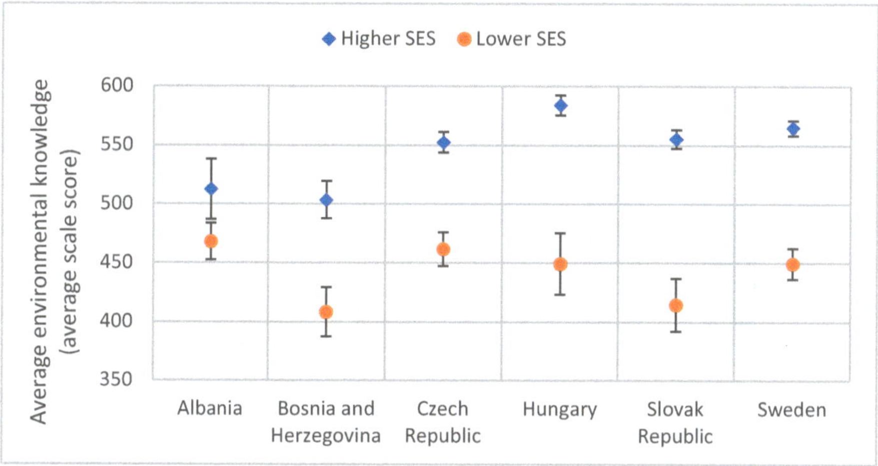

Fig. 7.3 Average environmental knowledge of grade four students from higher and lower socio-economic status groups. Note: *SES* Socioeconomic status

7.6 Results

This study aimed to identify patterns of knowledge and attitudes among students based on the TIMSS 2023 survey conducted in Albania, Austria, Bosnia and Herzegovina, the Czech Republic, Hungary, the Slovak Republic, and Sweden. There is a clear relationship between socioeconomic status and environmental knowledge among fourth-grade students across all analyzed countries (see Fig. 7.3). Students from higher socioeconomic backgrounds tend to exhibit greater levels of environmental knowledge, as reflected in their average scores.

In most countries analyzed, a relationship exists between socioeconomic status, knowledge, and pro-environmental values for fourth-grade students, as reflected in the Students Value Environmental Preservation scale. Higher socioeconomic status background and, consequently, higher levels of knowledge are associated with slightly higher values on this index. The exception is Bosnia and Herzegovina, where fourth graders from different socioeconomic groups show no variation in their Students Value Environmental Preservation scale scores.

A similar relationship exists between socioeconomic status and environmental knowledge among eighth graders in the countries analyzed: students from higher socioeconomic status backgrounds tend to demonstrate more significant levels of environmental knowledge. Additionally, there is a significant difference in pro-environmental values across different socioeconomic groups, as measured by the Students Value Environmental Preservation scale.

A notable trend observed exclusively among eighth graders in the countries analyzed is that a decline in pro-environmental values, as measured by the Students Value Environmental Preservation scale, corresponds to a significant decrease in

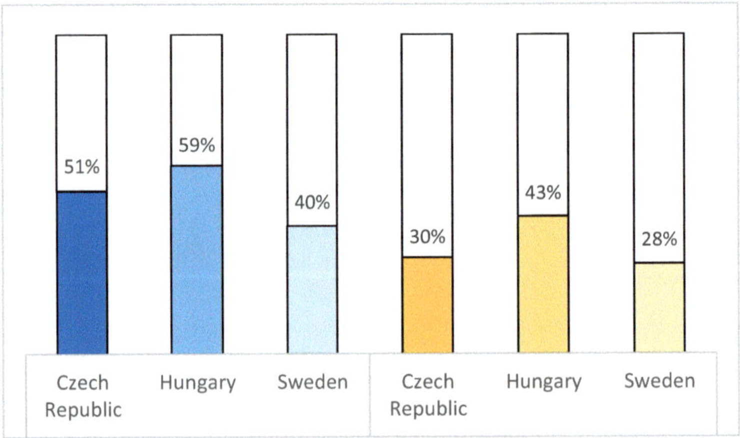

Fig. 7.4 Proportion of students reporting to *very strongly value* environmental preservation (grades four and eight)

their interest in spending time in nature. This pattern is not evident among fourth graders.

In every country included in the analysis that assessed fourth and eighth graders, the proportion of fourth graders demonstrating strong pro-environmental values (reported as *very strongly value*) was greater than that of eighth graders (see Fig. 7.4). Conversely, the proportion of fourth graders demonstrating low pro-environmental values (reported as *somewhat value*) was lower than that of eighth graders. Although the comparison of fourth and eighth graders is complicated due to groupings of the results on different cut-off points, it is important to mention so that more detailed analyses of changes in student values can be carried out in the future.

This study also aimed to identify patterns associated with community achievements in environmental sustainability—such as waste and wastewater management, climate action, biodiversity protection, and other factors reflected in the Sustainable Development Report—that could shape students' knowledge and attitudes.

The Sustainable Development Report 2024 (Sachs et al., 2024) covers 17 goals and their sub-indicators, ranging from poverty reduction to gender equality, which may dilute the focus on environmental-specific indicators. Focusing solely on environmental indicators, every country analyzed faces specific environmental challenges that may influence students' knowledge and pro-environmental attitudes.

However, the Sustainable Development Report reflects the wealth of individual countries in socioeconomic indicators, thus the socioeconomic status of the countries' population. When examining students' socioeconomic status in the analyzed countries, it becomes evident that Albania and Bosnia and Herzegovina differ significantly from the other four countries where the science and environmental knowledge survey of fourth graders was conducted. Therefore, significant differences in a country's ranking in the Sustainable Development Report can suggest lower levels of students' knowledge, as this analysis indicates that knowledge is closely linked to students' socioeconomic status. On the other hand, as the results from the countries analyzed show, fourth graders in countries with a lower ranking in the Sustainable Development Report and a higher proportion of students from lower socioeconomic backgrounds tend to have stronger pro-environmental values and a greater preference for spending time in nature. Other socioeconomic indicators or indices could serve as effectively as the Sustainable Development Report.

However, combining the Sustainable Development Report or its specific indicators with TIMSS can help pinpoint environmental issues not being addressed in a given country while also revealing gaps in students' knowledge needed to understand the problem or its solutions. In all the countries analyzed, major environmental problems are evident, and students show low success rates in solving related items.

7.6.1 Challenges in Air Pollution Items

Air pollution in urban areas is a challenge shared by all countries analyzed. However, fourth graders found it difficult to answer questions about reducing city air pollution. Albania had the highest success rate, with 34.9 percent of students giving correct answers. In comparison, around 20 percent of students in the Slovak Republic answered correctly, while Bosnia and Herzegovina recorded the lowest rate at just 13.1 percent. It should be noted here that the item dealing with air pollution reduction (Item ID SE81036) was classified as too difficult to anchor.

Eighth graders solved the item related to air pollution and its impact on animal health (Item ID SE82068B) and none of the analyzed countries saw a success rate above 50 percent. Students from the Czech Republic performed the best, with a success rate of 46.1 percent in this item, classified as advanced.

7.6.2 Challenges in Biodiversity Items

Biodiversity conservation is another significant challenge for many countries in this study, including Albania, Bosnia and Herzegovina, Hungary, and Sweden. Several TIMSS items related to ecosystems, organisms, the environment, and their interactions focus on biodiversity. However, students generally struggled with these items, particularly those involving food webs (Item ID SE81075). The best achievement in this item, classified as advanced, came from the Slovak Republic, where 26.1 percent of students provided correct answers.

Similarly, items requiring students to explain the relationship between environmental damage and its impact on organism populations (Item ID SE81054) also proved challenging. In this item, only Albania achieved nearly a third of the correct answers, while Sweden ranked second with just under a fifth of the correct answers. Due to the generally low number of correct answers, the item was considered too difficult to anchor.

Eighth graders also solved items related to food relationships within ecosystems. Most of the items were anchored at a low level, and the students completed them without any major difficulties. Eighth graders struggle with more complex items that require higher-level reasoning, such as understanding how changes in the populations of organisms that have a food relationship affect each other (Item ID SE82071). None of the countries achieved a success rate higher than 50 percent in this item, anchored as high, with students in Sweden performing the best at 47.8 percent correct answers.

7.6.3 Challenges in Climate Change Items

Climate change represents a significant challenge for many countries, including those analyzed in this study. Consequently, numerous TIMSS eighth grader items are dedicated to exploring this topic, focusing on its causes and effects on the natural environment and human society. Eighth graders' success in solving more complicated climate-related issues are below optimal levels. For instance, an item requiring students to demonstrate knowledge of the carbon cycle, specifically carbon sequestration, was completed most successfully by students in Sweden, with 34.5 percent answering correctly. In contrast, students from the Czech Republic performed the poorest among the analyzed countries, with only 21.2 percent correct answers. This item (Item ID SE72063) was considered too difficult to anchor.

There is also an interesting relationship between the average environmental knowledge scores of eighth graders and their perception of the urgency of climate change, its environmental and societal impacts, and the need to address these issues. In the two countries where eighth-grade students' average environmental knowledge scores matched or exceeded their overall science scores—Hungary and Sweden—a higher proportion strongly agreed that addressing climate change should be a high priority (see Fig. 7.5).

Country	Environmental Knowledge		Addressing climate change should be a high priority
	Average Scale Score	Difference from Overall Score	Percent of Students (Agree a Lot)
Sweden	534	▲	55%
Hungary	518		56%
Czechia	517	▼	49%
Austria	501	▼	46%

▲ ▼ Subscale score significantly different from overall score

▲ ▼ Subscale score significantly different from overall score

Fig. 7.5 Environmental knowledge and the priority of addressing climate change among eighth-grade students

The low success rate of students in the selected countries on items related to floods and their impact on drinking water supplies (Item ID SE82273A) is particularly notable. Eighth graders in the Czech Republic, for example, achieved only a 22.1 percent success rate on this item, which is anchored at the advanced level—40 percentage points lower than their peers in Sweden. This outcome is striking given that floods, as a consequence of climate change, are relatively frequent in the Czech Republic and often result in significant economic damage to affected areas. The poor performance of students from the Czech Republic on a topic so directly relevant to their local environment is both unexpected and puzzling.

7.6.4 Variations in Student Success in Difficult Environmental Items

The results presented reveal an intriguing phenomenon that warrants further investigation in the context of national curricula. Several items addressing environmental issues—topics of importance to many countries—are anchored at high, advanced, or even too difficult to anchor levels. In grade four, these items account for 62 percent of all environmental items, and in grade eight, they make up 68 percent. Items classified as too difficult to anchor make up nearly one fifth of all items in both grades.

Fourth graders struggle to identify examples of air pollution in cities (Item ID SE81036), the benefits of using sunlight or wind to generate electricity (Item ID SE71213), or why an increase in farmland area could lead to a decline in animal populations (Item ID SE81054). However, there are significant differences in success rates between countries. For example, on the item related to the benefits of using sunlight or wind to generate electricity (Item ID SE71213), the success rate among students from Hungary is 35.5 percentage points higher than that of their peers in Bosnia and Herzegovina.

Eighth-grade students in this study found it challenging to explain why temperatures are higher in city centers (Item ID SE72720), evaluate a conclusion about climate based on a single week of weather observations (Item ID SE62211B), or consider the carbon cycle in the context of carbon storage in wood (Item ID SE72063). Once again, significant differences in student success rates across countries were observed for certain items. For example, for an item dealing with the urban heat island effect (Item ID SE72720), the success rate of students from Austria was 18.9 percentage points higher than that of students from the Czech Republic. For an item dealing with the carbon cycle (Item ID SE72063), the success rate of students from Sweden was 12.3 percentage points higher than that of students from the Czech Republic.

These results prompt reflection on the observed differences and an exploration of their possible causes, which may be linked to the national or school curriculum or to teachers' approaches to teaching these topics.

7.7 Discussion and Conclusion

In the theoretical section of this chapter, one conception of worldview shaping decision-making was described as comprising five overlapping dimensions: ontological (fundamental assumptions about reality), epistemological (ways of acquiring and validating knowledge), axiological (core values), anthropological (views on human nature and humanity's role in the universe), and societal (how society is structured) (Hedlund-de Witt, 2012). These dimensions are closely related to how individuals and societies engage with sustainable development. This analysis explored the knowledge levels of fourth and eighth graders, their pro-environmental values, and their socioeconomic status. Moreover, the study highlighted the environmental challenges faced by the analyzed countries, emphasizing the critical link between worldview, decision-making, and sustainability.

This study revealed a relationship between socioeconomic status and level of environmental knowledge within the countries analyzed. Students from higher socioeconomic status backgrounds tend to demonstrate better environmental knowledge. This finding aligns with results from several other studies (for example, Buchmann, 2002; Finders et al., 2021; Lareau, 2011; Waters et al., 2021). Further, among the countries analyzed, a connection was observed between students' level of knowledge and their declared pro-environmental values. Higher knowledge scores were associated with stronger pro-environmental values. This finding is also consistent with results from other studies (for example, Chankrajang & Muttarak, 2017; Meyer, 2015).

However, while these patterns hold within a country, they may not apply across different countries. For example, fourth-grade students in Albania in the lower socioeconomic status group have an average environmental knowledge score of 468, but their Students Value Environmental Preservation scale score is 10.8. Only fourth-grade students in Hungary with an average environmental knowledge nearly 100 points higher reach a comparable value, and these students belong to a higher socioeconomic status group. Similarly, fourth-grade students in the Czech Republic and the Slovak Republic have identical Student Value Environmental Preservation scale scores of 9.9, but their average environmental knowledge scores are significantly higher, and these students also come from higher socioeconomic backgrounds. Thus, knowledge is not the sole factor influencing pro-environmental values. Based on this analysis, it can be hypothesized that the high pro-environmental values observed among students from Albania and Bosnia and Herzegovina may be attributed to the widespread appreciation of spending time in nature. However, several other factors may influence pro-environmental values, as highlighted in the study background, including, cultural perspectives (for example, human role in the universe), environmental degradation and its societal impact, and a societal emphasis on economic growth (see, for example, Post & Meng, 2018).

Each country has environmental indicators that highlight its environmental issues. This study found no direct relationship between students' knowledge, pro-environmental values, and the countries' rankings in the Sustainable

Development Report 2024. However, by combining environmental indicators with students' TIMSS scores, environmental topics where students may have gaps in understanding can be identified, and these could be addressed by emphasizing those topics in the curriculum.

This study showed that students can succeed in some items classified as difficult, as evidenced by the comparison of results across countries. These countries may serve as sources of inspiration regarding the intended or implemented curriculum. However, there are also items classified as too difficult to anchor, where students generally struggle to succeed. This presents an opportunity to explore how complex topics can be made more accessible to learners.

Furthermore, while knowledge alone does not guarantee the adoption of pro-environmental behavior, it undoubtedly plays a key role in making informed decisions, particularly when considering the potential impact of one's choices on the natural environment.

This study confirmed many of the findings from existing research and revealed new areas that warrant further exploration in environmental education. For example, it would be intriguing to investigate whether the pro-environmental values of eighth-grade students, compared to fourth-grade students, tend to decline as their focus shifts more toward finding their place in society. As students grow older, they may become more preoccupied with social and personal issues, potentially reducing their interest in environmental matters. Alternatively, it would be worth exploring whether a decline in pro-environmental values reflects the perceived complexity of translating these values into action. As students mature, they may recognize the challenges and limitations in addressing environmental issues, leading to a more nuanced or cynical view of the feasibility of meaningful environmental change. Understanding these dynamics could help shape more effective educational strategies that support the development of sustained pro-environmental values and behaviors at different stages of schooling.

Appendix 7.1: Key Environmental Indicators

		SDGs report assessment													
		Albania		Bosnia and Herzegovina		Czech Republic		Hungary		Slovak Republic		Austria		Sweden	
SDG	Indicator	State	Trend	State	State	State	Trend	State	Trend	State	Trend	State	Trend	State	Trend
6	Anthropogenic wastewater that receives treatment (%)	♦		♦	●	●		●		●		●		●	
11	Annual mean concentration of $PM_{2.5}$ (μg/m³)	□	↗	♦	□	□	↑	□	↑	□	↑	□	↑	●	↑
	Access to improved water source, piped (% of urban population)	■	↓	□	●	●	↑	●	↑	□	↑			□	→
12	Municipal solid waste (kg/capita/day)	□		●	●	●	↓	●	→	●		□	↓	●	→
13	CO_2 emissions from fossil fuel combustion and cement production (tCO_2/capita)	●	→	♦	♦	♦	→	♦	→	♦	→	♦	→	■	↗
14	Mean area that is protected in marine sites important to biodiversity (%)	■	→			Not applicable: country is landlocked.								♦	→
	Ocean Health Index: Clean Waters score (worst 0–100 best)	■	↑	●	↑									■	↑
	Fish caught from overexploited or collapsed stocks (% of total catch)													■	↓
	Fish caught by trawling or dredging (%)	♦	→	●										□	↑
	Fish caught that are then discarded (%)	♦	→	●	→									□	↑

15	Mean area that is protected in terrestrial sites important to biodiversity (%)	♦	→	♦	→	●	↑	□	→	●	↑	■	→	♦	→
	Mean area that is protected in freshwater sites important to biodiversity (%)	●	↑	●	↑	●	↑	●	↑	●	↑	■	→	♦	→
	Red List Index of species survival (worst 0–1 best)	●	→	□	→	●	→	●	→	●	→	□	→	●	↑
	Permanent deforestation (% of forest area, 3-year average)	●	↑	●	↑	●	→	●	↑	●	↑	●	↑	●	↑

Legend: ♦major challenges remain, ■ significant challenges remain, □ challenges remain, ● SDG achievement, data not available, ↓ decreasing, → stagnating, ↗ moderately increasing, ↑ on track

Source: Sustainable Development Report 2024 (Sachs et al., 2024)

Appendix 7.2: Anchor Level of TIMSS 2023 Environmental Items, Grade Four

Item ID	Content domain	Topic area	Anchor level
SE61015	Life science	Ecosystems	High
SE61069	Life science	Ecosystems	Intermediate
SE71063	Life science	Ecosystems	High
SE71065	Life science	Ecosystems	High
SE71069	Life science	Ecosystems	Advanced
SE71071	Life science	Ecosystems	Advanced
SE71076	Life science	Ecosystems	Intermediate
SE71080	Life science	Ecosystems	Advanced
SE71081	Life science	Ecosystems	Advanced
SE81042	Life science	Ecosystems	Low
SE81070	Life science	Ecosystems	Intermediate
SE81073	Life science	Ecosystems	High
SE81075	Life science	Ecosystems	Advanced
SE81077	Life science	Ecosystems	Intermediate
SE81079	Life science	Ecosystems	Advanced
SE81080	Life science	Ecosystems	Low
SE81083	Life science	Ecosystems	Intermediate
SE81085	Life science	Ecosystems	Intermediate (1 of 2 points)
SE81085	Life science	Ecosystems	High (2 of 2 points)
SE81036	Life science	Organisms, environment, and their interactions	Too difficult to anchor
SE81044	Life science	Organisms, environment, and their interactions	Low
SE81054	Life science	Organisms, environment, and their interactions	Too difficult to anchor
SE81055	Life science	Organisms, environment, and their interactions	High
SE81076	Life science	Organisms, environment, and their interactions	Advanced
SE81168	Life science	Organisms, environment, and their interactions	Intermediate
SE81207A	Physical science	Forms of energy and energy transfer	Intermediate
SE81207B	Physical science	Forms of energy and energy transfer	Low
SE61124	Earth science	Earth's physical characteristics, resources, and history	Too difficult to anchor
SE71201	Earth science	Earth's physical characteristics, resources, and history	Intermediate
SE71213	Earth science	Earth's physical characteristics, resources, and history	Too difficult to anchor

Item ID	Content domain	Topic area	Anchor level
SE71214	Earth science	Earth's physical characteristics, resources, and history	High
SE71220	Earth science	Earth's physical characteristics, resources, and history	Advanced
SE71222	Earth science	Earth's physical characteristics, resources, and history	High
SE71223	Earth science	Earth's physical characteristics, resources, and history	Low
SE81200	Earth science	Earth's physical characteristics, resources, and history	High
SE81202	Earth science	Earth's physical characteristics, resources, and history	Low
SE81204	Earth science	Earth's physical characteristics, resources, and history	Too difficult to anchor
SE81212	Earth science	Earth's physical characteristics, resources, and history	Advanced
SE81920A	Earth science	Earth's physical characteristics, resources, and history	Advanced
SE81920B	Earth science	Earth's physical characteristics, resources, and history	Advanced
SE61116	Earth science	Earth's weather and climates	High
SE71237	Earth science	Earth's weather and climates	High
SE81238	Earth science	Earth's weather and climates	Low
SE81239	Earth science	Earth's weather and climates	Low

Appendix 7.3: Anchor Level of TIMSS 2023 Environmental Items, Grade Eight

Item ID	Content domain	Topic area	Anchor level
SE62089	Biology	Ecosystems	Advanced
SE62091A	Biology	Ecosystems	Intermediate
SE62091B	Biology	Ecosystems	Intermediate
SE72063	Biology	Ecosystems	Too difficult to anchor
SE72066	Biology	Ecosystems	High
SE72074	Biology	Ecosystems	High
SE72082	Biology	Ecosystems	High
SE72086	Biology	Ecosystems	Low
SE72460	Biology	Ecosystems	High
SE82060	Biology	Ecosystems	Low (1 of 2 points)
SE82060	Biology	Ecosystems	High (2 of 2 points)
SE82064	Biology	Ecosystems	Intermediate

Item ID	Content domain	Topic area	Anchor level
SE82068A	Biology	Ecosystems	High
SE82068B	Biology	Ecosystems	Advanced
SE82070	Biology	Ecosystems	Low
SE82071	Biology	Ecosystems	High
SE82075	Biology	Ecosystems	High
SE82077	Biology	Ecosystems	High
SE82080	Biology	Ecosystems	Intermediate
SE82273A	Biology	Ecosystems	Advanced
SE82273B	Biology	Ecosystems	Advanced
SE82902	Biology	Ecosystems	High
SQ82L01	Biology	Ecosystems	Intermediate
SQ82L02A	Biology	Ecosystems	Low
SQ82L02B	Biology	Ecosystems	Low
SQ82L04A	Biology	Ecosystems	Low
SQ82L04B	Biology	Ecosystems	High
SQ82L07A	Biology	Ecosystems	Intermediate
SQ82L07B	Biology	Ecosystems	Intermediate
SQ82L09A	Biology	Ecosystems	Intermediate
SQ82L09B	Biology	Ecosystems	Advanced
SE62177	Earth science	Earth's processes, cycles, and history	Advanced
SE62180	Earth science	Earth's processes, cycles, and history	High
SE62211A	Earth science	Earth's processes, cycles, and history	High
SE62211B	Earth science	Earth's processes, cycles, and history	Too difficult to anchor
SE62243	Earth science	Earth's processes, cycles, and history	Intermediate (1 of 2 points)
SE62243	Earth science	Earth's processes, cycles, and history	High (2 of 2 points)
SE72323	Earth science	Earth's processes, cycles, and history	High
SE72720	Earth science	Earth's processes, cycles, and history	Too difficult to anchor
SE82323	Earth science	Earth's processes, cycles, and history	Too difficult to anchor
SE82325	Earth science	Earth's processes, cycles, and history	Too difficult to anchor
SE82327	Earth science	Earth's processes, cycles, and history	High
SE82331	Earth science	Earth's processes, cycles, and history	Intermediate
SE82333	Earth science	Earth's processes, cycles, and history	High
SE82335	Earth science	Earth's processes, cycles, and history	Advanced
SE82721	Earth science	Earth's processes, cycles, and history	Intermediate
SE62190	Earth science	Earth's resources, their use and conservation	High
SE62235	Earth science	Earth's resources, their use and conservation	Advanced
SE72345	Earth science	Earth's resources, their use and conservation	Intermediate (1 of 2 points)
SE72345	Earth science	Earth's resources, their use and conservation	Too difficult to anchor (2 of 2 points)

Item ID	Content domain	Topic area	Anchor level
SE72348	Earth science	Earth's resources, their use and conservation	Low
SE72349	Earth science	Earth's resources, their use and conservation	High
SE82099	Earth science	Earth's resources, their use and conservation	Advanced
SE82340	Earth science	Earth's resources, their use and conservation	Intermediate
SE82341	Earth science	Earth's resources, their use and conservation	Too difficult to anchor
SE82343	Earth science	Earth's resources, their use and conservation	High
SE82346A	Earth science	Earth's resources, their use and conservation	High
SE82346B	Earth science	Earth's resources, their use and conservation	High
SE82347	Earth science	Earth's resources, their use and conservation	Too difficult to anchor
SE62022	Earth science	Earth's structure and physical features	Advanced
SE82233	Earth science	Earth's structure and physical features	High

References

Ajzen, I. (1991). The theory of planned behavior. *Organizational Behavior and Human Decision Processes, 50*(2), 179–211. https://doi.org/10.1016/0749-5978(91)90020-T

Berenguer, J., Corraliza, J. A., & Martín, R. (2005). Rural-urban differences in environmental concern, attitudes, and actions. *European Journal of Psychological Assessment, 21*(2), 128–138. https://doi.org/10.1027/1015-5759.21.2.128

Bolisani, E., & Bratianu, C. (2018). The elusive definition of knowledge. In E. Bolisani & C. Bratianu (Eds.), *Emergent knowledge strategies: Strategic thinking in knowledge management* (pp. 1–22). Springer. https://doi.org/10.1007/978-3-319-60657-6_1

Buchmann, C. (2002). Measuring family background in international studies of education: Conceptual issues and methodological challenges. In A. Porter & A. Gamoran (Eds.), *Methodological advances in cross-national surveys of educational achievement* (pp. 150–197). National Academy Press. https://doi.org/10.17226/10322

Cenia. (2008). *Hospodářství a životní prostředí v České republice po roce 1989* [Economy and environment in the Czech Republic after 1989]. https://www.cenia.cz/wp-content/uploads/2019/04/Hospodarstvi_a_ZP_CR_po_roce_1989.pdf

Cenia. (2022). *Zpráva o životním prostředí České republiky 2021* [Environmental report of the Czech Republic 2021]. https://www.cenia.cz/wp-content/uploads/2023/04/Zprava_ZPCR_2021.pdf

Centurino, V. A. S., & Kelly, D. L. (2021). TIMSS 2023 science framework. In I. V. S. Mullis, M. O. Martin, & M. von Davier (Eds.), *TIMSS 2023 assessment frameworks*. TIMSS & PIRLS International Study Center, Boston College. https://timssandpirls.bc.edu/timss2023/frameworks/chapter-2.html

Chankrajang, T., & Muttarak, R. (2017). Green returns to education: Does schooling contribute to pro-environmental behaviours? Evidence from Thailand. *Ecological Economics, 131*, 434–448. https://doi.org/10.1016/j.ecolecon.2016.09.015

Chwialkowska, A., Bhatti, W. A., & Glowik, M. (2020). The influence of cultural values on pro-environmental behavior. *Journal of Cleaner Production, 268*, 122305. https://doi.org/10.1016/j.jclepro.2020.122305

CIA. (2024a). *The World Factbook. Albania.* https://www.cia.gov/the-world-factbook/countries/albania/

CIA. (2024b). *The World Factbook. Bosnia and Herzegovina.* https://www.cia.gov/the-world-factbook/countries/bosnia-and-herzegovina/

CIA. (2024c). *The World Factbook. Czechia.* https://www.cia.gov/the-world-factbook/countries/czechia/

CIA. (2024d). *The World Factbook. Hungary.* https://www.cia.gov/the-world-factbook/countries/hungary/

CIA. (2024e). *The World Factbook. Slovakia.* https://www.cia.gov/the-world-factbook/countries/slovakia/

CIA. (2024f). *The World Factbook. Austria.* https://www.cia.gov/the-world-factbook/countries/austria/

CIA. (2024g). *The World Factbook. Sweden.* https://www.cia.gov/the-world-factbook/countries/sweden/

Csákvári, E., Fabók, V., Bartha, S., Barta, Z., Batáry, P., Borics, G., Botta-Dukát, Z., Erős, T., Gáspár, J., Hideg, E., Kovács-Hostyánszki, A., Sramkó, G., Standovár, T., Lengyel, S., Liker, A., Magura, T., Márton, A., Molnár, A. V., Molnár, Z., et al. (2021). Conservation biology research priorities for 2050: A Central-Eastern European perspective. *Biological Conservation, 264*, 109396. https://doi.org/10.1016/j.biocon.2021.109396

De Groot, J. I., & Steg, L. (2008). Value orientations to explain beliefs related to environmental significant behavior: How to measure egoistic, altruistic, and biospheric value orientations. *Environment and Behavior, 40*(3), 330–354. https://doi.org/10.1177/0013916506297831

de Vries, B. J. M. (2019). Engaging with the Sustainable Development Goals by going beyond modernity: An ethical evaluation within a worldview framework. *Global Sustainability, 2*, e18. https://doi.org/10.1017/sus.2019.15

Eurofish. (2023). *Albania.* https://eurofish.dk/member-countries/albania/

European Environment Agency. (2022). *Early warning assessment related to the 2025 targets for municipal waste and packaging waste.* https://www.eea.europa.eu/en/analysis/publications/many-eu-member-states/early-warning-assessment-related-to-the-2025-targets-for-municipal-waste-and-packaging-waste

European Union. (2023). *Eurostat report shows EU's progress on Sustainable Development Goals.* https://ec.europa.eu/commission/presscorner/detail/en/ip_23_2887

Finders, J. K., McClelland, M. M., Geldhof, G. J., Rothwell, D. W., & Hatfield, B. E. (2021). Explaining achievement gaps in kindergarten and third grade: The role of self-regulation and executive function skills. *Early Childhood Research Quarterly, 54*, 72–85. https://doi.org/10.1016/j.ecresq.2020.07.008

Franzen, A., & Vogl, D. (2013). Two decades of measuring environmental attitudes: A comparative analysis of 33 countries. *Global Environmental Change, 23*(5), 1001–1008. https://doi.org/10.1016/j.gloenvcha.2013.03.009

Garner, B. (2017). "Perfectly positioned": The blurring of urban, suburban, and rural boundaries in a southern community. *The Annals of the American Academy of Political and Social Science, 672*(1), 46–63. https://doi.org/10.1177/0002716217710490

Gilchrist, A. (2013). Design for living: The challenge of sustainable communities. In H. Barton (Ed.), *Sustainable communities* (pp. 147–159). Routledge. https://doi.org/10.4324/9781315870649

Goula, M., Ladias, C. A., Gioti-Papadaki, O., & Hasanagas, N. (2015). The spatial dimension of environment-related attitudes: Does urban or rural origin matter? *Regional Science Inquiry, 7*(2), 115–129.

Government of Hungary. (2023). *National biodiversity strategy for 2030*. https://www.cbd.int/doc/world/hu/hu-nbsap-v3-en.pdf

Hedlund-de Witt, A. (2012). Exploring worldviews and their relationships to sustainable lifestyles: Towards a new conceptual and methodological approach. *Ecological Economics, 84*, 74–83. https://doi.org/10.1016/j.ecolecon.2012.09.009

Ignatow, G. (2006). Cultural models of nature and society: Reconsidering environmental attitudes and concern. *Environment and Behavior, 38*(4), 441–461. https://doi.org/10.1177/0013916505280791

International Trade Administration. (2024). *Bosnia and Herzegovina country commercial guide*. https://www.trade.gov/country-commercial-guides/bosnia-and-herzegovina-agriculture

Janoušková, S., Teplý, P., Fatka, D., Teplá, M., Cajthaml, T., & Hák, T. (2020). Microplastics—How and what do university students know about the emerging environmental sustainability issue? *Sustainability, 12*(21), 9220. https://doi.org/10.3390/su12219220

Koltko-Rivera, M. E. (2004). The psychology of worldviews. *Review of General Psychology, 8*(1), 3–58. https://doi.org/10.1037/1089-2680.8.1.3

Lareau, A. (2011). *Unequal childhoods: Class, race, and family life* (2nd ed.). University of California Press. https://www.jstor.org/stable/10.1525/j.ctt1ppgj4

Lou, X., Li, L. M. W., & Ito, K. (2024). Egoistic value is positively associated with pro-environmental attitude and behaviour when the environmental problems are psychologically close. *British Journal of Social Psychology, 63*(4), 1640–1657. https://doi.org/10.1111/bjso.12743

Martin, C., & Czellar, S. (2017). Where do biospheric values come from? A connectedness to nature perspective. *Journal of Environmental Psychology, 52*, 56–68. https://doi.org/10.1016/j.jenvp.2017.04.009

Meyer, A. (2015). Does education increase pro-environmental behavior? Evidence from Europe. *Ecological Economics, 116*, 108–121. https://doi.org/10.1016/j.ecolecon.2015.04.018

Naturvardsverket. (2024). *A rich diversity of plant and animal life*. https://www.naturvardsverket.se/en/om-miljoarbetet/swedish-environmental-objectives/a-rich-diversity-of-plant-and-animal-life/

OECD. (2025). *Environment at a glance. Indicators*. OECD Publishing. https://doi.org/10.1787/ac4b8b89-en

Our World in Data. (2024). *SDG tracker: Measuring progress towards the Sustainable Development Goals*. https://ourworldindata.org/sdgs

Palacios Madero, M. D., Torío López, S., & Murga-Menoyo, M. Á. (2023). Conductas pro-sostenibilidad y percepciones del estudiantado adolescente sobre prácticas parentales y valores familiares [Pro-sustainability behaviors and perceptions of adolescent students about parenting practices and family values]. *Interdisciplinaria Revista de Psicología y Ciencias Afines, 40*(2), 409–423.

Pe'er, S., Goldman, D., & Yavetz, B. (2007). Environmental literacy in teacher training: Attitudes, knowledge, and environmental behavior of beginning students. *The Journal of Environmental Education, 39*(1), 45–59. https://doi.org/10.3200/JOEE.39.1.45-59

Post, D., & Meng, Y. (2018). Does schooling foster environmental values and action? A cross-national study of priorities and behaviors. *International Journal of Educational Development, 60*, 10–18. https://doi.org/10.1016/j.ijedudev.2017.10.010

Quimby, C. C., & Angelique, H. (2011). Identifying barriers and catalysts to fostering pro-environmental behavior: Opportunities and challenges for community psychology. *American Journal of Community Psychology, 47*(3–4), 388–396. https://doi.org/10.1007/s10464-010-9389-7

Reynolds, K. A., & Komakhidze, M. (2022). *TIMSS 2023 environmental attitudes and behaviors framework*. TIMSS & PIRLS International Study Center, Boston College. https://

timssandpirls.bc.edu/timss2023/frameworks/pdf/T23-Environmental-Attitudes-and-Behaviors-Framework.pdf

Richardson, K., Steffen, W., Lucht, W., Bendtsen, J., Cornell, S. E., Donges, J. F., Drüke, M., Fetzer, I., Bala, G., von Bloh, W., Feulner, G., Fiedler, S., Gerten, D., Gleeson, T., Hofmann, M., Huiskamp, W., Kummu, M., Mohan, C., Nogués-Bravo, D., et al. (2023). Earth beyond six of nine planetary boundaries. *Science Advances, 9*(37), eadh2458. https://doi.org/10.1126/sciadv.adh2458

Rigolot, C. (2018). Sustainability transformations as shifts in worldviews. *Ecology and Society, 23*(2). https://doi.org/10.5751/ES-10101-230222

Rockström, J., Steffen, W., Noone, K., Persson, Å., Chapin, F. S., III, Lambin, E., Lenton, T., Scheffer, M., Folke, C., Schellnhuber, H., Nykvist, B., De Wit, C. A., Hughes, T., van der Leeuw, S., Rodhe, H., Sörlin, S., Snyder, P. K., Costanza, R., Svedin, U., et al. (2009). Planetary boundaries: Exploring the safe operating space for humanity. *Ecology and Society, 14*(2), 32. http://www.ecologyandsociety.org/vol14/iss2/art32/

Rokeach, M. (1968). *Beliefs, attitudes, and values*. Jossey-Bass.

Rokeach, M. (1973). *The nature of human values*. Free Press.

Rosida, R. F., Amaliah, L. N., Mahardika, I. K., & Suratno, S. (2023). The process of forming knowledge: In the study of ontology, epistemology, and axiology. *International Journal for Educational and Vocational Studies, 5*(1), 13–18. https://doi.org/10.29103/ijevs.v5i1.12980

Sachs, J. D., Lafortune, G., & Fuller, G. (2024). *The SDGs and the UN Summit of the Future. Sustainable Development Report 2024*. Dublin University Press. https://doi.org/10.25546/108572

Schultz, P. (2002). Environmental attitudes and behaviors across cultures. *Online Readings in Psychology and Culture, 8*(1), 4. https://doi.org/10.9707/2307-0919.1070

Schwartz, S. H., & Bilsky, W. (1987). Toward a universal psychological structure of human values. *Journal of Personality and Social Psychology, 53*(3), 550–562. https://doi.org/10.1037/0022-3514.53.3.550

Snelgar, R. S. (2006). Egoistic, altruistic, and biospheric environmental concerns: Measurement and structure. *Journal of Environmental Psychology, 26*(2), 87–99. https://doi.org/10.1016/j.jenvp.2006.06.003

Steg, L., & De Groot, J. I. (2012). Environmental values. In S. D. Clayton (Ed.), *The Oxford handbook of environmental and conservation psychology* (pp. 81–92). Oxford University Press. https://doi.org/10.1093/oxfordhb/9780199733026.013.0005

Swigger, N. (2013). The online citizen: Is social media changing citizens' beliefs about democratic values? *Political Behavior, 35*(3), 589–603. https://doi.org/10.1007/s11109-012-9208-y

Taber, K. S. (2019). *The nature of the chemical concept: Re-constructing chemical knowledge in teaching and learning* (Vol. 3). Royal Society of Chemistry.

United Nations. (2015). *Transforming our world: The 2030 Agenda for Sustainable Development* (A/RES/70/1). https://sdgs.un.org/2030agenda

United Nations. (2017). *Work of the Statistical Commission pertaining to the 2030 Agenda for Sustainable Development: Resolution adopted by the General Assembly on 6 July 2017* (A/RES/71/313). https://documents.un.org/doc/undoc/gen/n17/207/63/pdf/n1720763.pdf

van Egmond, N. D., & de Vries, H. J. M. (2011). Sustainability: The search for the integral worldview. *Futures, 43*(8), 853–867. https://doi.org/10.1016/j.futures.2011.05.027

Van Opstal, M., & Hugé, J. (2013). Knowledge for sustainable development: A worldviews perspective. *Environment, Development and Sustainability, 15*, 687–709. https://doi.org/10.1007/s10668-012-9401-5

Vejsiu, I. N. A., & Stafasani, M. (2017). Biodiversity effect in ecotourism. *Albanian Journal of Agricultural Sciences*, 271–276.

von Davier, M., Fishbein, B., & Kennedy, A. (Eds.). (2024). *TIMSS 2023 technical report (methods and procedures)*. TIMSS & PIRLS International Study Center, Boston College. https://timss2023.org/methods

von Davier, M., Kennedy, A., Reynolds, K., Fishbein, B., Khorramdel, L., Aldrich, C., Bookbinder, A., Bezirhan, U., & Yin, L. (2024). *TIMSS 2023 international results in mathematics and science*. TIMSS & PIRLS International Study Center, Boston College. https://doi.org/10.6017/lse.tpisc.timss.rs6460

Waters, N. E., Ahmed, S. F., Tang, S., Morrison, F. J., & Davis-Kean, P. E. (2021). Pathways from socioeconomic status to early academic achievement: The role of specific executive functions. *Early Childhood Research Quarterly, 54*, 321–331. https://doi.org/10.1016/j.ecresq.2020.09.008

Weitz, N., Persson, Å., Nilsson, M., & Tenggren, S. (2015). *Sustainable development goals for Sweden: Insights on setting a national agenda*. Stockholm Environment Institute.

Wiseman, M., & Bogner, F. X. (2003). A higher-order model of ecological values and its relationship to personality. *Personality and Individual Differences, 34*(5), 783–794. https://doi.org/10.1016/S0191-8869(02)00071-5

Withalm, G., & Getzner, M. (2018). *Protected areas and regional development: An Austrian case study*. InTech. https://doi.org/10.5772/intechopen.72373

Yu, X. (2014). Is environment 'a city thing' in China? Rural–urban differences in environmental attitudes. *Journal of Environmental Psychology, 38*, 39–48. https://doi.org/10.1016/j.jenvp.2013.12.009

Svatava Janoušková is a full professor at the Faculty of Science (Charles University). She has participated in the development and revisions of national science curriculum and is the co-author of several science textbooks in the Czech Republic.

Her long-term research interests include the theory and application of sustainability and educational indicators. Internationally, she collaborates on several research projects including TIMSS.

Paulína Koršňáková is a Senior Research and Liaison Adviser to IEA, and facilitates research networking, stakeholder engagement, and collaboration that enhance capacity building and knowledge sharing to foster innovation and quality in education. She serves as a series editor for the IEA Research for Educators and IEA Teacher Snippets. She holds a master degree in Teacher Preparation Programme of Biology and Mathematics and a doctorate in general psychology.

Libor Klement works at the headquarters of the Czech School Inspectorate in Prague as an expert in international large-scale assessments. Since 2021, he has held the position of national coordinator for the TIMSS study at both the fourth and eighth grade levels. His research focuses on the Czech education system and its international comparison.

Michaela Janoušková is a psychology student with a strong interest in the research of human values and motivation. She focuses particularly on environmental values and sustainable behavior. She is interested in how values influence everyday decision-making and people's attitudes toward environmental protection.

Michal Blaško works at the Czech School Inspectorate in the Department of International Studies. He is involved in TIMSS and PISA studies. He currently works in a team revising the national curriculum for science subjects and has long been involved in integrating science content into the curriculum at both the national and school levels.

Pavel Zikl is associate professor at the Faculty of Education (University of Hradec Králové) and head of the Institute of Primary, Pre-primary and Special Education.

His research focuses on the education of pupils with special educational needs, inclusive education and the specifics of education of pupils with special educational needs in different areas of the curriculum.

MIX
Papier aus verantwortungsvollen Quellen
Paper from responsible sources
FSC® C105338

If you have any concerns about our products,
you can contact us on
ProductSafety@springernature.com

In case Publisher is established outside the EU,
the EU authorized representative is:
Springer Nature Customer Service Center GmbH
Europaplatz 3, 69115 Heidelberg, Germany

Printed by Libri Plureos GmbH
in Hamburg, Germany